BEYOND THE
RANGELAND CONFLICT

"*Beyond the Rangeland Conflict* reveals the power of conservation/ranching partnerships, and shows that when debate becomes dialogue, innovation results. Its message will discomfort many conservationists and ranchers alike—which is exactly why it should be read by everyone concerned with the future of biodiversity on our western rangelands."

JOHN C. SAWHILL, CEO and President, The Nature Conservancy

"Solutions, not conflict, that's what *Beyond the Rangeland Conflict* gives us, in a wonderfully readable format: ten ranches, each one unique and fascinating, each one managed in a loving, restorative, and sustainable way. This book will cure the western land use blues."

GRETEL EHRLICH, author of *The Solace of Open Spaces* and *A Match to the Heart*

"*Beyond the Rangeland Conflict* is as readable as it is inspiring, and the people in these pages are as hardy as the land they love. They need to be, because of how hard it will be to create a sustainable way of inhabiting that land. Dan Dagget's stories will spread hard-earned inspiration and hope from one end of the West to the other."

DANIEL KEMMIS, Mayor of Missoula, Montana, and author of *Community and the Politics of Place*

"Grasslands must be grazed, or they become something other than grasslands. Ranchers and environmentalists have far more compatible goals than they do with many other would-be users of our rangelands. *Beyond the Rangeland Conflict* succinctly describes those goals and gives vivid examples where the concerned and reasonable people have successfully cooperated to improve and preserve our grazing lands."

ROBERT M. MILLER DVM, columnist for *Western Horseman*

BEYOND THE
RANGELAND

TOWARD A

CONFLICT

WEST THAT WORKS

by Dan Dagget with portraits by Jay Dusard

GIBBS·SMITH
PUBLISHER

IN COOPERATION WITH
THE GRAND CANYON TRUST

ACKNOWLEDGMENTS

THESE THOUGHTS CAME TO ME as I was driving across Nevada, headed home from visiting one of the ranches described in this book. I was passing through an immense valley, in basin-and-range country, where the dramatic features of the topography were made even more so by the warm light and the stark shadows of sunset. The scene was positively breathtaking. It was one of those moments when you feel that you have been touched directly by the universe, and what has been revealed is beyond words. I got a hitch in my throat and gave thanks to whatever spirits were listening. They have afforded me the opportunity to get to know some of the most remarkable lands on this small and intimate planet and to be enriched by some of the most exceptional people I will ever meet—people who manifest a dedication so deep, abiding, and true that Nature makes the land green around them.

With that in mind, I would like to mention some of those dedicated people who were brought to my attention but could not be given a chapter in this book. This was entirely because of limits on space and time and not because of the quality of their stewardship.

They include: Doc and Connie Hatfield, Hatfield's High Desert Ranch, Oregon; Dwight and Isleta Cooper, the Hayhook Ranch, Arizona; Dayton, Gerda, John, and Taylor Hyde, the Yamsi Ranch, Oregon; Les and Linda Davis, the CS Ranch, New Mexico; Dennis and Deborah Moroney, M.D., the Cross-U Ranch, Arizona; and Doc Roberts, the Crooked H Ranch, also in Arizona.

For the generous support they gave to this project, The Grand Canyon Trust, Jay Dusard, and I would like to express our thanks to: The National Fish and Wildlife Foundation, The Horace W. Goldsmith Foundation, The Margaret T. Morris Foundation, The Pacificorp Foundation, The Rocky Mountain Elk Foundation, The Transition Foundation, J. W. Kieckhefer Foundation, Coleman Natural Meats, and Common Ground.

I would like to express my deep gratitude to Tommie Martin, without whom this book would not exist nor would many of the success stories therein.

Last, but certainly not least, I would like to thank my wife, Trish Jahnke, for the patience and support she gave during a project that was at times all-consuming and seemed destined to go on forever.

First Edition

97 96 95   5 4 3 2 1

Copyright © 1995 by The Grand Canyon Trust

This book is copublished by
The Grand Canyon Trust, "The Homestead,"
Route 4, Box 718, Flagstaff, Arizona 86001
and
Gibbs Smith, Publisher,
P.O. Box 667, Layton, Utah 84041

Half-title photo: Rangeland, Tipton Ranch, Nevada
Frontispiece photo: Rangeland, Gray Ranch, New Mexico

Editor: Nicky Leach
Book design: Larry Lindahl
Project coordinator for the Grand Canyon Trust: Rick Moore
Coordinator for Gibbs Smith, Publisher: Dawn Valentine Hadlock
Black & white photography, and sketches: © 1995 Jay Dusard
Color photography: © 1995 Dan Dagget except pages 19, 22, 54, 82-83
   by Jay Dusard, and page 43 courtesy Rancho de la Inmaculada

Library of Congress Cataloging-in-Publication Data

Dagget, Dan, 1944-
    Beyond the rangeland conflict : toward a West that works / written
    by Dan Dagget in cooperation with the Grand Canyon Trust :
    photographs by Jay Dusard.
        p.   cm.
    ISBN 0-87905-654-1 (pbk.) : $19.95
    1. Range management—West (U.S.) 2. Grazing—Environmental
aspects—West (U.S.) 3. Range ecology—West (U.S.) I. Dusard, Jay.
II. Grand Canyon Trust. III. Title.
SF85.35.W4D34 1995
333.74´153´0978—dc20   95-11727
                                                                              CIP

Printed in the United States of America

CONTENTS

"If we are to correct our abuses of each other and of other races and of our land, and if our effort to correct these abuses is to be more than a political fad that will in the long run be only another form of abuse, then we are going to have to go far beyond public protest and political action. We are going to have to rebuild the substance and the integrity of private life in this country. We are going to have to gather up the fragments of knowledge and responsibility that we have parceled out to the bureaus and the corporations and the specialists, and we are going to have to put those fragments back together again in our own minds and in our families and households and neighborhoods."

WENDELL BERRY
from Continuous Harmony: Essays Cultural and Agricultural

FOREWORD

ENVIRONMENTALISTS MAY BE DIVIDED by Dan Dagget's book. Some will lament that the book does not support arguments that cows should be removed from every acre of public lands. Instead, it tells the stories of quiet, uncelebrated efforts of a few ranchers around the West who changed their grazing practices and reversed the loss of riparian habitat, biodiversity, and recreational opportunities on public lands.

Ranchers may be of two minds as well. Those who insist that all is well on the range and are committed to business as usual will see all this as unnecessary and potentially costly advice from outsiders. Yet some ranchers who have controlled large blocks of the West for long periods now realize that many of those lands could be healthier, more attractive, and more productive than they now are.

The debate over grazing has become polarized. Therefore, stories of ranchers trying something new and different that improves the quality of public land, while keeping cattle—sometimes *more* cattle—on it, may be confusing or disappointing to people who are drawn to the simplicity of absolutes or extremes.

We do not offer this book as the "position" of the Grand Canyon Trust. It is not. It is Dan Dagget's anecdotal account of what he saw, heard, and felt on ten ranches around the West, where ranchers are experimenting with unconventional grazing techniques designed to be sensitive to the peculiar needs of their lands—and where success is being measured in improved wildlife habitat, as well as better economic returns.

Although the Trust claims no special expertise in range science, we do see a problem with public land grazing, and we have a sense that it is serious. We read federal reports that say 158 million acres of public rangelands (59 percent) are in "poor" or "fair" condition. The National Science Foundation, in its 1994 report, found it impossible to make sound scientific judgements about the conditions and trends of public rangelands, despairing over the lack of reliable data. We *do* know, however, of some degraded, nearly lifeless lands around the Colorado Plateau, and we see them assaulted year after year by livestock. It will take more than anecdotes about successful grazing experiments elsewhere to convince us that the fragile soils and delicate habitat of many of the dry, sparsely vegetated areas we know on the plateau could ever thrive as grazing lands. We have our own anecdotes. For instance, some of us have been impressed by the recovery of streamside vegetation in Grand Gulch since the cattle were moved out. We compare it to Comb Wash, where stream banks denuded by cattle became vulnerable to scouring floods and the vegetation now struggles to recover, even though grazing has been limited by a court injunction.

Still, through our work as advocates for conserving the resources of the Colorado Plateau, we have come to know and respect communities and individuals around the region that grew up ranching—and loving—the same lands we do. We've discovered that neighbors don't have to agree on everything to work together when there is some common ground. It's not hard to find ranchers on the plateau who share some of our most heartfelt values. Most want their grandchildren to know a region with healthy ecosystems and places of wonder, beauty, and solitude. And most can understand that economic stability and permanence of communities are intertwined with the permanent health of the surrounding land, water, and wildlife.

We are committed to sitting down with others around the West who start with similar core values to try to find solutions that work for the lands we cherish. So it seems only fitting to tell the stories of ten ranchers who invited their neighbors, "experts," environmentalists, and others to work with them to find better ways to manage their rangelands. They speak with pride of revegetated lands, larger and more diverse wildlife populations—and higher profits. Many of their fellow ranchers remain dubious and critical of both the methods used and the need for them. Range scientists question what the experiments "prove." And, yes, we still have a hard time visualizing Comb Wash flourishing with a herd of cattle in it! But the grazing debate has been stalled too long by recalcitrance, closed minds, and a lack of innovation. If these ranchers dared to try and find a better way, we believe that their stories deserve to be heard.

DAVID H. GETCHES, *Chairman*
Board of Trustees, Grand Canyon Trust
April 1995

INTRODUCTION

A FTER READING an advance copy of this book, a friend of mine, whose opinion I respect very much, said to me: "One thing that worries me about what you've written is that I know someone's going to hold it up and wave it around and say it proves that everything's all right out there on the range; that the case people all the way back to John Muir have been making about the problems ranching has caused in the West has been trumped up, or at least wildly exaggerated. That's a problem."

And I had to agree with him. There will be people who will do just that. And this will cause problems for those of us who are working very hard to find political solutions to the very real environmental crisis that exists on America's rangelands. But I believe that a greater danger exists in not telling these stories, in letting political considerations silence the hope born of the genuine efforts of dedicated people who have succeeded at least to a degree in living in harmony with the land.

Since I share my friend's concern about the misuse of these stories, I decided to state it plainly and clearly, up front here, where it's not hard to find: Anyone using this book to support the status quo is misrepresenting it, is perpetrating an intentional deception. While the stories included here are about people coming together to heal the damaged ecosystems of North America's rangelands, an essential part of that story is that one of the reasons those lands need to be healed at all is because of the abuses many of those same people, or their forebears, brought to it.

Far from being free of problems, I believe that much of the western range is in worse shape than even some of the most alarming assessments would have us believe. We've been told, and we believe, that many of the West's mountains, valleys, prairies, and deserts are denuded and eroding; that most of its streams have been robbed of the water that is their lifeblood till they are nothing more than storm sewers periodically gutted by flash floods roaring down from barren uplands; that the wildlife community that once lived here in a way we suspect was sustainable has itself become a government dependent, stripped of the predators and other natural forces that once held it in check, so that it requires extensive "management" to moderate its boom-and-bust excesses. These things we know and find alarming enough. But what many of us don't know is that a significant portion of these magnificent and irreplaceable lands have deteriorated to the point where they are no longer able to rebound. And, of these, even the ones that have been left for nature to heal are getting worse instead of better. I tell you this not because I've read it in a book or a government report but because I've seen it.

During the six years that I researched this book (three of them before I knew I was going to write it), I drove and flew tens of thousands of miles, from Sonora, Mexico, to Montana, and back again. I hiked, mountain biked, and horsebacked across ranches, parks, and preserves that cover millions of acres. I stooped with monitoring form and loupe in hand until my back ached and my hamstrings cramped. I scrambled up perpendicular rock faces to mesa tops where livestock have never grazed, and I had people take me out onto the land to show me what they believed was irrefutable evidence that I didn't have the slightest idea what I was talking about.

In these travels, I visited areas that proved as plainly as it can be proven that our way of relating to the arid open lands of the West is a failure. In these wastelands, most of which historic photographs and other evidence suggest have been created since the arrival of European culture, the bare dirt that separates living plants is measured in yards instead of inches. The desiccated carcasses of the plants that once inhabited those interstices litter the lifeless soil. The roots of the few survivors that persist dangle hopelessly in the raw wounds of fresh-cut gullies like skeletal fingers clutching at empty space in a final, futile effort to hold back soil that is relentlessly washing away. The land here is quite simply dead, and nature is removing it to give it a chance to support life somewhere else.

While I have been in these places, not one rancher, bureaucrat, or apologist has said that things were as they should have been. Nor has anyone said that livestock grazing is innocent of having caused this devastation. But a few have gone beyond the obvious to make an observation that is more revealing, more valuable than whining the same old complaints. It is an observation that has the sound of a wake-up call, of the kind of realization that can be the harbinger of change. They pointed out that, as much as this devastation is the result of more than a century of grazing mismanagement, it is also the legacy of more than a hundred years of failed remedies.

It was in the early 1890s that environmentalist John Muir succeeded in having cattle and sheep banned from the infant collection of forest preserves that was to become our national forest system. In short order Muir's success was overturned, setting off a see-saw battle between the proponents of preservation and the "wise use" of rangelands that persists today. Since neither side has been able to score a decisive victory in this war, its history has been a string of stiff doses of contention followed by the bitter aftertaste of compromise. Limits have been put on the number of cattle that can be grazed on public lands, and those limits have been reduced, reduced, and

reduced again. Fees have been set for the use of those lands, and we have argued nonstop over whether or not to increase those fees. Billions of dollars of government money have been spent to build fences and water sources to disperse animals and spread their impact, and millions of acres have had to be bulldozed clean of one round of mistakes and reseeded only to be ready for the blade again in less than a generation. As a final expression of frustration, livestock has been excluded entirely from some places. Today, these places provide us with an opportunity to see the results of the very solutions that most of us are still clamoring for. Some of them date back more than seventy years. Surprisingly, more often than not, it is difficult to tell which side of the fence has been cured.

Frustrated by solutions that yield the same results as the problems they're designed to remedy, a few individuals from both sides of this standoff have left the meeting rooms for the landscape, rolled up their shirtsleeves, and gone to work. In some cases, they've stepped into the traces side by side. They will be the first to tell you it hasn't been easy.

There have been times (and I've been around for a few) when it was all that even extremely tolerant people could do just to stay within earshot of one another. In some cases, activists on the environmentalist side have refused to enter into these collaborations or (if they have entered into them) have thrown up their hands and left, saying it's not even worth trying; that working with adversaries outside the hearing room or the courtroom causes a watering down of resolve on the part of once-stalwart eco-warriors and leads to compromises that stand in the way of real solutions.

Those who have stuck with the process say they have done so because of the successes they are achieving on the ground—successes they measure in terms of green, growing grass, and increased biodiversity, and soil staying where it was formed rather than washing away. But even these successes are not above question. Some scientists say they are more the result of a misinterpretation of the data than of real change—more a mirage of false optimism than an accumulation of real biomass. Others say that they're the result of higher-than-usual rainfall, or that they're merely temporary and won't last. As I've walked the land with some of those who have invested their blood, sweat, and tears in these efforts and watched as they became excited over changes I could barely see, I've certainly asked those same questions. But every time I was just about convinced that this was all a chimera, and that the progress that some good steward had made was too little for a challenge so immense, or that it would have happened just as well without his or her intervention, that's when I invariably came across something that was nothing less than amazing—something that restored my confidence and renewed my commitment.

The critics are right at least to the degree that there is room in all these instances for a healthy skepticism. (Isn't that always the case?) Most of the projects discussed in this book have been underway for less than ten years. But while the debate continues among academics and activists, the "applied scientists" whose stories are profiled here are taking people out on the land to show them their side of the story. I went and looked, and now I'm ready to tell you what I saw.

DAN DAGGET
Flagstaff, Arizona
February 1995

"Rather than trying to freeze the West in mythic time, Wise Users
ought to be joining environmentalists in an earnest search for both livelihood
and community....Clearly it's time to try something new."

DONALD SNOW
from The Pristine Silence of Leaving It All Alone

Storms have been nourishing diversity on the western range for millennia. Here, wildflowers
and grass catch the light under a darkening sky in western Montana.

CHAPTER ONE

THE
WESTERN RANGE

*On Nose Studs and Common Ground; Issues, Issues, and
More Issues; Feral Oranges in Montana and Longhorns in Texas;
Megafauna; Mine Sites; and What Have We Got to Win?*

THIS BOOK isn't just about ranches and grasslands, although that certainly is its focus. If you read it with only that in mind, you will miss much of what it has to say, and much of what has been achieved by the truly exceptional people described within these pages. What these stories are about is the potential that exists for solving some of our most pressing environmental problems if we define success in terms of achieving goals on the land rather than winning out over one another.

While working on this book, I was invited to speak to an ecology class at a private secondary school near Sedona, Arizona. The instructor asked me to spend an hour or so with her students to present to them what she called a broader view on rangeland issues. The Verde Valley School, which takes a liberal approach to education and has an experimental emphasis, is located in Arizona's Red Rock Country, at the southern rim of the Colorado Plateau, in an area that has come to be as famous for contemporary "hip" and New Age culture as it is for spectacular scenery. The school looks out on a scene of towering cliffs and desert grassland, some of which I helped get designated as wilderness as a volunteer with the Sierra Club in 1984.

The teacher who invited me said her class had been studying range issues for two-and-a-half weeks. For a textbook, they had been using Lynn Jacobs's encyclopedic indictment of western public lands ranching, *Waste The West*. Jacobs's position on ranching is simple: He's against it 100 percent ("...Any kind of ranching is significantly more harmful than non-ranching," he writes). The teacher told me she had heard that

I had a slide show of ranches that weren't all bare dirt and cow pies and an interesting talk to go with it. With a little encouragement from one of her students, whose mother was a sheepherder, she had decided to give her class a look at more than a monoculture of perspectives on the matter.

When I arrived, I found a dozen teenagers, some with dreadlocks and nose rings, others wearing skateboard shorts and backwards hats. All were obviously as bored at the prospect of sitting through a program on environmentally benign ranching as they would have been listening to a medley of Lawrence Welk's greatest hits. As I set up the projector and talked about my efforts at bringing environmentalists and ranchers together, they informed me they'd already done a conflict resolution simulation on the issue and had come up with their own common ground solution. I asked what it was, and they told me.

In return for not totally revoking the privilege of ranchers to graze on public lands, they said, they would quadruple grazing fees, fence live-stock out of all streamside areas, make less public land available for livestock in general, and mandate lower densities on what was available. In addition, they would create large numbers of ungrazed comparison plots throughout the grazed lands to show ranchers just how much damage they were causing.

I couldn't help but smile at their "common ground" solution. It would have put most ranchers I know into cardiac arrest, or on a Wise Use movement contributors list.

I told them that I go about finding common ground in a different way, that I start by having people of diverse, even opposite points of view identify the goals each of them wants to achieve on the land. Then I encourage those apparent adversaries to work together to reach whatever of those goals they hold in common. With that in mind, I asked the class what they would like the rangelands of the West to be like. They began listing the things we all want: more green plants and less bare dirt, clear-flowing streams, healthy riparian areas, more wildlife. I was writing all this on the blackboard when a young woman stopped me.

"Why are we wasting our time with this?" she asked impatiently. "We're naming all this pie-in-the-sky stuff, and we know it's never going to be like that, because we're never going to be able to get rid of all of them."

I said maybe it was time to take a look at the slide show.

The first few images were what everyone expected: landscapes stripped bare of everything that wasn't too tough or too prickly to eat. Everyone sat there, arms crossed. They'd seen it before. Then came a photograph that created enough of a stir that even those who were sleeping woke up. It showed a riparian area with grasses and rushes and saplings of cottonwood and willow bordering a clear stream that was almost lost among the greenery. The vegetation was so lush it looked unreal, but it was real all right.

"Where's that?" several of the students exclaimed in honest surprise.

"Phil Knight's Date Creek Ranch, not very far south of here," I answered. "There were 500 cattle in this very place two months before I took this photograph. It's grazed five months out of every year."

As we continued through the slide show, the class got to see all the goals they had listed on the blackboard brought to life on ranches from Mexico to the Canadian border. At the end of the session their mood had changed dramatically. They were enthusiastic. They asked questions. They patted me on the back. One wanted to know if I could get her a job on such a ranch.

I've shown those slides to a number of groups, using a variety of different presentations. I've even set them up to compare photographs taken on ranches with photographs shot on nearby nature preserves—on the same day—in similar areas. Then I ask the audience if they can tell the ranches from the preserves. Invariably, they pick the wrong ones, until they figure out that the green ones are the ranches.

The point that I try to make with this pre-sentation is not that all livestock grazing is good; nor am I saying that the people who express deep concern about the damage ranching has caused throughout the West are mistaken—my photographs show plenty of ranches with over-grazed uplands and denuded streambanks. And I'm not saying that every place on any of these ranches looks better than a park or preserve. They don't. What I am saying is that the trend on these well-managed ranches is toward more biodiversity and biomass rather than less; that significant por-tions of them are in good to excellent condition (no matter what your criteria for "good" is); and that the places that aren't are getting better.

Furthermore, I'm saying that all this is the way it is because of the manner in which these lands are being managed.

I'm aware of the case that's been made against livestock grazing in the West. For nearly ten years, as conservation chair of the Northern Arizona Sierra Club group, and as a writer for a number of environmental journals, including the *Earth First! Journal*, I helped make that case. I added my voice to the chorus that said much of the West is too dry, or too fragile, or too irreplaceable for grazing; that the plants in the Southwest, the Colorado Plateau, and the Great Basin evolved without large natural grazers, and so they can't be asked to support large exotic ones now; that livestock are the main reason behind the extirpation of the wolf, jaguar, and grizzly, and that cattle have contributed to the near extermination of the black-footed ferret and the persecution of other creatures, from prairie dogs to pronghorns, too numerous to mention.

I've repeated—too many times to count—the statistics that make up the indictment against western ranchers: that domesticated grazers are responsible for destroying up to 90 percent of some western states' riparian areas that are, in turn, vital to up to 80 percent of the region's indigenous species; that livestock are the reason 59 percent of our public rangelands are in poor condition; that the belches of livestock contribute to global warming, their excrement fouls our campgrounds and pollutes our streams, and bits of their bodies clog our arteries.

And I still believe that—at least most of it. But I also believe things are changing. The way ranchers ranch (some of them, at least) is changing. The way we think about the land and each other is changing.

In 1989, I worked to repeal an Arizona law that made it legal for ranchers to kill mountain lions and bears, essentially at the ranchers' discretion. We circulated petitions. Some of us engaged in demonstrations. The confrontations kept getting hotter and more contentious, with no sign of a letup, until a couple of forward-looking individuals from each camp invited some of the most outspoken environmentalists and the most middle-of-the-road ranchers to meet face to face. The intent was to see if such an encounter would drop the megatonnage of word bombs being lobbed back and forth. Twelve of us showed up. We were evenly split: six and six.

At that first meeting, we were sincerely surprised that we could talk to one another without having the conversation degenerate into a shouting match—and just as surprised that we shared some of the same values: a love for open spaces, a hitch in our chests when we saw wildlife, and a concern over population growth and the inexorable march of suburbs. After that first meeting we kept getting together—at first, I believe, out of pure amazement that we could do it without ending up at each other's throats. After all, our group included members of Earth First!, the Sierra Club, and Preserve Arizona's Wolves, along with a couple of mountain lion hunters, the president-elect of the Arizona Cattlegrowers Association, and ranchers who had never talked to an environmentalist outside a hearing room. We began to call ourselves "6-6" and said it stood for six of us and six of them, even though our numbers grew to several times that.

Eventually, we decided it didn't make sense for people interested in rangelands to meet in suburban living rooms, so we started getting together at member's ranches. There, we found ourselves talking more about the land and less about other things. There, the idea for this book was conceived and began gestating.

Spending time in such a diverse group, on the very landscape we were all so concerned about, we began to see that the problems on the western range aren't as simple as the media have been painting them. As time went along, it became more and more apparent that there were no villains here, nor heroes—just people. And the land is so much more complex when you're standing on it than it appears in a magazine photograph with an inflammatory caption underneath it. If you want to make it more complex yet, pick up a range monitoring data sheet and learn how to use it. I did, and I started to branch out beyond the ranches of the members of 6-6.

After looking at hundreds of thousands of acres of rangeland and hearing of the work scores of people had done, not only to stop degrading the land but to restore and improve it, I found my perspective on how we can best live as a positive force on America's rangelands changing. I found myself beginning to believe that we have been focusing too much on what is wrong with one another and not enough on what each of us has to offer; too much on how one side or the other can win the fight and too little on how we can all make things better; too much on issues and too little on the land.

Those realizations were no more than amorphous feelings, however, until I asked a leader of a regional environmental group what I should be looking for in deciding whether or not a rancher was a good steward. She responded without hesitation: Do they support ecosystem management? What about exotic plants: are they trying to get rid of them in favor of native plants? Have they cut livestock numbers? Have they fenced them out of riparian areas? How do they stand on wolf reintroduction? On the reintroduction of wildfire? Do they use federal Animal Damage Control agents to kill predators? She kept right on rattling off a long list of issues. Issues. And more issues. But she never said even once what the land should look like.

That's when the light went on. That's when I came to believe that, in the environmental debate, issues have ceased being the means and have come to be the end. She—we—are prepared to decide whether or not a person is a good manager without ever looking at their land, and we're willing to decide whether or not the land is healthy without ever seeing it. (How many of us would know how to decide if we did see it?) If that land is "right" in terms of the issues, and if it's protected by a legal fence of the proper regulations, we assume it must be healthy. If not, some scoundrel must be subverting those regulations or some slob must be running roughshod over them. And if a piece of land is not right with the issues, then it goes without saying that it's in bad shape and getting worse. We make those assumptions every time we write a letter to a land management agency saying, "Do this." "Don't do that."

By relating to ecosystems as if they were collections of issues, we turn natural landscapes into political ones. And by so doing, we bludgeon a magnificently diverse world into a war zone of opposites: right and wrong, us and them, winners and losers. We demand blind loyalty of all those involved in these issues, dismissing as "sold out" anyone who changes their position for whatever reason—even a good one. Issues turn us into confrontation junkies. They pit us against one another instead of against our problems. Under their influence our goal becomes defeating the enemy rather than improving, restoring, and reviving our damaged ecosystems. And if either side ever does win, it will find its reward to be a

hollow one, since in the very act of winning it has created a losing side that vows to become stronger and take its turn as victor next time.

Issues create another problem. They are based on the assumption that what is needed is for those of us on the right side to tell those of us on the wrong side what to do. But does someone in Key West, Florida, or Washington, D.C., or even in Phoenix, Arizona, really know enough about rangelands and grasses in, say, Sonoita, Arizona, to tell a rancher there how to manage his land? Do you, wherever you live, know how to make the grass grow greener and thicker in central Nevada on less than ten inches of annual rainfall? Maybe not, you're probably saying, but you know what you like: green grass and flowing streams. And you know what you don't like: overgrazed meadows and streams full of excrement.

If you are now thinking, Don't tell them what to do; tell them what we want, we've both arrived at what the people whose stories are included in this book have come to believe is the best way out of this dilemma. Instead of telling one another what to do, perhaps we should all be saying what we want. For one thing, no one likes to be told what to do; it just makes them resist, and we've seen plenty of resistance. For another, many of us don't know enough about rangeland ecosystems, and certainly not enough about specific ranches, to tell someone how to manage them. The truth is that the experts most of us rely on—scientists, politicians, and activists—don't know either. They know how to win political battles and how to get you to send them money, but most of them are operating on assumptions about rangeland ecology that were made in the nineteenth century. Unfortunately, so are many ranchers. Though most of us know very little about the ecology of rangelands, all of us know what we like, and what we want. And though none of us likes to be told what to do, most of us like to be challenged by goals, and we like to get credit for attaining them when we do. The most exciting part of all this is, just as those students at the Verde Valley School discovered, when it comes to the land, most of us—ranchers and environmentalists alike—want the same things.

That's all well and good, some of you will say, but doesn't our responsibility here boil down to doing what it takes to heal the land, even if that means leaving it alone? And haven't we admitted that cows are what caused the land to need to be healed in the first place? So, how can ranchers improve the land and keep on ranching? Isn't that like hoping to cure lung cancer while continuing to smoke cigarettes? The way to restore the land is to take the cows off, right? And if a few ranchers did make the land better, that's how they did it, right? If cows are bad we've already said that it stands to reason that less cows must be better (the same goes for sheep, horses, and goats), and no cows must be best. It's like the old joke:

"Doc, it hurts when I do this."

"Then, stop doing it."

But one thing twenty years of working as an activist has taught me is that anything that looks simple almost never is; that just about the time you think you've got it all figured out you run into something that shatters your smugness.

While there have been plenty of cases where taking cows off the land has yielded results that people were happy with, there have also been a disturbing number of instances, especially with land that has been severely degraded, where a healthy dose of this cure-all hasn't worked—and an intriguing number of cases (especially the worst ones) where tossing it out the door has.

On the Jornada Experimental Range north of Las Cruces, New Mexico, cattle and sheep were excluded seventy-five years ago from study plots where nearly three centuries of grazing had replaced the land's cover of grasses with scattered creosote bush and bare dirt. Recovery was so slow on those lands that, after twenty-five years, jackrabbits were excluded from some of them to see if that would speed the healing. Fifty years after that the expected regeneration has yet to occur. "The grasses haven't come back," Jornada director Kris Havstad told a reporter for the *Albuquerque Journal*, concluding that "just taking the animals off the land won't restore it." This hasn't happened only on the Jornada, it's been observed in hundreds, maybe thousands, of exclosures all over the West.

Numerous explanations have been offered for this perplexing state of affairs—from "we must be in a dry cycle" to "grazing by cattle caused the land to lose critical gasses." Some say that grazing has caused such severe damage to the fragile ecosystems of the arid West that some lands may never recover. "Massive intervention" (meaning bulldozers and extensive seeding) is talked of as the only hope to restore the devastated ecosystems. But while some continue their bedside vigil, waiting for signs of recovery in a patient that has failed to respond to three quarters of a century of convalescence, a team of ranchers and environmentalists, whose story is part of this book, managed to bring life to a barren, topsoil-less mine site in central Nevada in a single year, when only six inches of moisture fell. And they didn't do it by resting the land, or by massive intervention with heavy machinery and fossil fuels. They did it in a way that the conventional wisdom is absolutely at a loss to explain—they used the very agents we consider the destroyers of vegetation: cows.

Other cases in this book pose a similar challenge to our conventional way of thinking. In almost every success story related here, the number of grazing animals on the land wasn't decreased, it was increased. In some cases it was more than doubled, and in one case it was quadrupled. Besides being effective, these methods of restoration generated cash flow as they healed, offering rural western communities ways to restore the land that won't send them begging to the federal dole.

This is a hard pill to swallow, especially for people who have walked through moonscapes dotted with cow splats. George Wuerthner, one of my favorite "no compromise" environmental writers, maintains that bringing cows that evolved in the moist meadows of Europe to the arid American West makes as little sense and requires as much subsidization and manipulation as raising oranges in Montana. Though I have yet to hear of anyone rounding up feral citrus in the Big Sky State, the great cattle drives that spread across the West and spawned ranches as far north as Canada, were, for the most part, relocations of hordes of wild Spanish cattle that proliferated *au naturel* in the arid grasslands of southwestern Texas.

Cattle survived and even proliferated in the American West because they were able to exploit niches that were vacant not only of bison but also of other large grazers and browsers that lived here during the Pleistocene Era more than 10,000 years ago. At that time, paleoecologists tell us, the savannas of North America and Eurasia made up one of the most prolific habitats the world has ever known. In the fall of 1992, on a field trip to a ranch and a preserve in southeastern Arizona, Tony Burgess, a noted grassland ecologist from the University of Arizona, was asked how one could justify putting exotic animals such as cattle on a North American grassland. Burgess replied, "The question we should be asking is 'Are they exotic enough?'" Burgess went on to explain that the area in which we were standing was typical habitat for a subspecies of mammoth. The time has been so short since those large grass eaters became extinct, he declared, that, in terms of evolution, "The plants don't know they're gone."

Ecologists studying the ancient grazers that existed in North America during the great ice ages report evidence of an interdependency, if not a direct symbiosis, between ungulates and the grassland habitat that supported them. In his book, *Frozen Faunas of the Mammoth Steppes*, Dayle Guthrie, professor of zoology at the Institute of Arctic Biology at the University of Alaska, tells us that not only were the animals of the savannas that once stretched from the British Isles to Mexico adapted to the grasses they grazed but that the reverse was also true. Guthrie lists dispersal of seeds, fertilization of soil, and removal of dead tissue, which exposed plants and soil to the sun's warming rays, as "favors" large grazers performed for the plants of the Pleistocene savannas.

S. J. McNaughton, of the Research Biology Lab at Syracuse University, has studied grasslands and the way they respond to grazing by wild ungulates around the world. McNaughton cautions that his studies, which have been centered on the Serengeti Plain of Africa and in Yellowstone National Park in the United States, have dealt mostly with the natural grazing of free-roaming wildlife. In view of that, McNaughton says his results may not apply to ranching and livestock. "Human management, even the most intelligent and enlightened, is not as effectual . . . as are the mechanisms produced by fifty million years of evolution," he states.

Nevertheless, McNaughton's research shows a relationship between wild grazers and grass very similar to the one that people who are included in this book are working to achieve with livestock. Ivan Aguirre moves his herd of 3,000 cattle over the desert grasslands of northern Mexico in a manner designed to have an effect similar to the migrating wildebeest McNaughton has studied on Africa's Serengeti. Aguirre reports results similar to those McNaughton described in an article in the February 1993 issue of *Ecological Applications*, when the latter wrote, ". . . moderate grazing promotes the productivity of many grasslands above the levels that prevail in the absence of grazing."

As for the effects of dense herds of animals such as those grazed by Aguirre and many of the other ranchers in this book, McNaughton wrote, "dense herding behavior allows them [grazers] to crop the grasslands in a way that increases forage yield."

Alan Savory, a land management consultant who learned about grazing while working as a wildlife biologist in Zimbabwe, has developed a management planning model based upon interactions similar to those described by Guthrie, McNaughton, and others. While Savory's Holistic Resource Management (HRM) model remains extremely controversial among ranchers as well as environmentalists, a growing number of ranchers use it, or their own adaptation of it, to chart a course in managing their lands. Others have developed their own management systems based on the principles of HRM: that time is more important than numbers with regard to the effect grazing animals can have on the land; that excluding all types of disturbance from grasslands (grazing, fire, trampling) can actually have a detrimental effect upon them; and that grazing animals can be used as a tool to improve the land, especially if they're herded into dense concentrations that mimic the effects of herds of wild grazers.

No matter what environmentalists think of Alan Savory or the planning model and management concepts he has developed, it is hard to escape the fact that the ranchers who use HRM, or some adaptation of it, not only speak the lingo of environmentalism but they manage for many of the same goals that environmentalists value: biodiversity, healthy riparian areas, and sustainability. It should come as no surprise, then, that a large proportion of the ranchers included in this book use HRM or some part of it.

If it is possible to graze domestic animals upon the lands of the West in ways that no longer destroy but that sustain and even restore, there are more reasons for urban and rural environmentalists to work together than there are to keep battling. If common ground exists upon which these two old adversaries can work together and achieve what both of them want, then to continue a fight that has persisted virtually unchanged for more than a century is either an expression of insanity, a mistake, or an admission there is something more at issue here than the environment.

Humans are now inextricably involved in the

management of virtually every inch of land on the planet. Even designating land as wilderness is, after all, a form of management. When we do so, we set up strict controls as to what people can and cannot do on that land. We say that they can take a hike on it, but they can't ride a bike on it; that they can bring a camera or a fishing reel, but not a hang glider. When you tell people that they can do some things on a piece of land but not do others, you're managing it. When you pass a law that says wilderness must be treated one way, but not another, you're managing it. Ironically, by designating land as wilderness, we even preclude the reintroduction into it of some of the forces that made it what it is today: the use of fire, the hunting of animals year-round, and even the cultivation of land by the people who evolved with it and helped shape its ecosystems.

Humans have been part of the ecosystems of the western hemisphere for at least 11,000 years, and perhaps more than 30,000. During that time, ecologists tell us that our species has become inextricably woven into the web of interactions we now call nature. Appropriately, for our purposes here, the habitat type in the West in which Homo sapiens has had the greatest impact is the grassland-savanna type. Some ecologists now tell us that, since the extinction of the Pleistocene grazers, human-caused disturbance has been responsible for the very existence of grasslands in North America. Stephen Pyne, in his book, *Fire in America*, tells how American Indians used fire to maintain and even expand the grasslands that were home to the great herds of bison that existed when Europeans first arrived here. By that time, Pyne writes, the Buffalo Indians had become so accomplished at husbanding (although he doesn't use that word) the herds with natural processes such as fire, herding, and predation that the West was supporting large grazing animals in the nearly unbelievable numbers that harked back to the Pleistocene.

Larry Agenbroad, a paleoecologist specializing in Quaternary Studies at Northern Arizona University, travels around the West in search of ancient Native American buffalo slaughter sites. For thousands of years before the horse returned to the hemisphere via Spanish galleons, indigenous peoples stampeded herds of bison over low cliffs to kill the animals in sufficient numbers to feed their villages. Agenbroad tells of one area of cliffs in southern Idaho that was an intricately

designed collection of stone barriers, hiding stations, jumps, holding pens, and butchering sites used by indigenous hunters for thousands of years. This "processing site" reveals, he says, that ranching, in one form or another, has been going on in the West for millennia.

It is by reviving those old relationships, as nearly as possible in the context of the contemporary West, that the ranchers in this book are trying to restore the land's natural vitality and diversity. While some of us say that to do so with cattle instead of buffalo makes this inevitably a sham, to "reintroduce" the bison herds without the Stone Age hunters that managed them would most likely do little more than create an immense overgrazed zoo.

The main objective of this book is to chronicle the success stories of these ranchers, and, as much as possible, the management teams with whom they work to increase biodiversity, revive riparian areas and watersheds, and restore the vitality of grasslands and savannas. By recognizing the efforts of ranchers who have successfully stewarded the land, we hope to encourage some of their colleagues to make their case with good management rather than confrontation. By using the words of good stewards to tell how they have benefitted from working with teams of individuals who otherwise may have been their adversaries, we hope to inspire other ranchers to form such relationships. By telling of the successes some of these collaborations have realized in ecosystem restoration, we hope to encourage more environmentalists to work with ranchers and find their reward on the land, rather than in the hearing room or the courtroom—even if they do it from across the country by helping a team set a goal. And finally, by describing some of the techniques involved in causing these groups to succeed, we hope to provide sufficient information to make it easier for those who wish to follow in their footsteps to do so.

"It is not necessarily those lands which are the most fertile or most favored in climate that seem to me the happiest, but those in which a long struggle or adaptation between man and his environment has brought out the best qualities of both."

T. S. ELIOT
from After Strange Gods

GRAY RANCH, NEW MEXICO

Front row (left to right): Ed Roos, Bill Miller, Ray Turner (seated), Don Dwyer (seated), Drum Hadley, Seth Hadley.

Second row (left to right): Ross Humphreys, Larry Allen, Tom Peterson, Wendy Glenn, Warner Glenn, Mary McDonald, Bill McDonald.

Back row (seated, left to right): Mike Dennis, Ron Bemis, Ben Brown, Joe Austin, Billy Darnell, John Cook.

CHAPTER TWO

THE
GRAY RANCH

Drummond Hadley and the Animas Foundation

NEW MEXICO

On Preserving Open Space, Bringing Together Diverse Interests,
and Reintroducing Fire Into the Ecosystem

O N AN APRIL MORNING in New Mexico's southwestern boot heel, dawn's pale glow reveals three figures huddled under a huge, gnarled cottonwood, their shoulders scrunched against the chill. The Mexican border is only a few miles south, but the elevation here on this sprawling savanna at the foot of the Animas Mountains is 5,000 feet. Today, that's high enough to make the weather more like Montana than Acapulco.

"Might as well start a fire," says one of the figures, dressed in a canvas jacket, brown Stetson, and bat-wing chaps. "It looks like we got here a little early. Seth might be awhile." The three of us rummage around, picking up dead branches that have fallen off the cottonwood and piling them on top of a tumbleweed stomped into kindling. Finally, someone strikes a match to it. The tumbleweed flares with a whoosh, and the twigs catch fire. Everyone gets as close as they dare, sizzling on one side and freezing on the other.

Our host is Drummond Hadley—rancher, poet, visionary. Hadley is waiting for the final closing of escrow on his purchase of the Gray Ranch from the Nature Conservancy. It's a deal that has left many people in the environmental community scratching their heads. The ranch, one of the most beautiful and uniquely diverse stretches of open country still intact in the Lower Forty-eight, was acquired by the Conservancy in 1990 and became part of an ambitious campaign, named Last Great Places, to save the best of what was left of the planet. At the time, the purchase was described as the largest single private conservation acquisition in the history of the United States, perhaps in the world. Now, four years later, the Conservancy was selling the Gray back to a cattle rancher. Some will be asking why for a long time.

I had come to the ranch with photographer Jay Dusard to learn about this apparently backwards deal and about this very special place and what the future holds for it. To Drum Hadley, getting to know a piece of open country means literally getting a feel for it on horseback, preferably as a matter of work rather than idle observation. When we arrived the night before, he let us know that our horses would be ready before sunup. We were to ride with him and his son Seth, Seth's compadre Kendra, and Richard and David Moore, to gather cows and calves for the following day's branding. We accepted gladly. Jay has done quite a bit of cowboying in his time, and I've always loved horses.

Back at the fire, we finally begin to warm up enough to talk about something other than the cold, and Hadley tells us that he bought the 502-square-mile Gray from the Conservancy because of his dedication to keeping as much of the West as possible open and undeveloped. Another reason was that his neighbors would rather have him own it than any of the others to whom the Conservancy was considering selling. "And at my age it's good to feel useful to somebody," he jokes

with a deep-creased smile. Hadley's face is ruddy from many seasons in the sun. His shock of thick, dark brown hair and beard are just beginning to show streaks of gray. His voice is slow and measured. It cracks a bit, like a reedy saxophone, as he continues.

"Maintaining a grazing livelihood on private lands like these is the only way to keep them and the surrounding public lands as open space," Hadley says, mouthing each of these words as if building one of his poems (or a testament). The heart of the Gray—the Animas Mountains and the Animas Valley—is within the ranch's 226,000 acres of private land. The other 326,000 acres are public land, mostly BLM. "If ranchers can't make a living at grazing, they sell off their private lands and most likely those lands are subdivided. Then the government starts trading off the surrounding public lands. That nearly always means more people," he concludes, "and the American West is further impoverished of its once-limitless open space."

For Hadley to be speaking of the Gray Ranch as a spearhead in the effort to solidify a future for ranching constitutes what is, at least on the surface, an immense turnabout. The Conservancy's purchase of the Gray was seen by everyone, from local cowboys who feared it to

Getting to know a piece of open country means literally getting a feel for it on horseback, preferably as a matter of work rather than idle observation.

national conservation groups who hailed it, as the first step in a logical sequence of events to turn the ranch into a huge preserve. The Department of the Interior had even picked a name for that preserve: the Animas National Wildlife Refuge.

Whether rancher or conservationist, anyone intent on saving open space, especially the most worthy examples of it, would have to place the Gray near the top of the list. Named for a

"'Y'know, Drum, you used to be able to count the cedar trees you could see from here. Now there are thousands beyond counting.' Walter Ramsey told me that while we was standing right here," said Drum Hadley.

Texas Ranger who came but did not stay, the Gray was consolidated from a collection of homesteads during the last two decades of the nineteenth century. The man who put it together was George Hearst, father of newspaper magnate William Randolph Hearst.

Because water was never developed here to the extent it has been on nearby ranches, and because the mountains never yielded a promise of great mineral riches, the Gray has never suffered the insults of overgrazing and mining that have devastated so much of the West. Upon seeing its broad grasslands and pine-clad mountains with their columnar palisades, virtually the same as they have been for centuries, Nature Conservancy President John Sawhill is reported to have said, "This may be one of the most beautiful places on earth."

The Conservancy's database of endangered species and unique natural communities, the most extensive in the world, confirms that the Gray is as biologically irreplaceable as it is beautiful. Split by the Continental Divide, which runs along the backbone of the Animas Mountains, and bordered on the south by Mexico, the ranch is home to

sixty distinct plant communities. One-third of those are described as rare or threatened. Within this one-of-a-kind mosaic of habitats live more than a thousand species of plants, seventy-five species and subspecies of mammals, and fifty-two species of reptiles and amphibians—at least that's what was found as of the last count. Add 150 species of birds that nest on the ranch, and an endangered species tally that includes seven mammals, three reptiles, two amphibians, and fifteen birds, and the result is a concentration of diversity that defies comparison to any other like-sized piece of North America.

And the ranch is big—big enough to inspire journalists to go thumbing through their dog-eared reference books for attention-grabbing comparisons to make its immensity comprehensible to urban readers. "Half the size of Rhode Island," touts one. "Big enough to hold the cities of Baltimore, Denver, Paris, Vienna, Buenos Aires and Cairo—if you could deal with the parking," quips another. Both of those writers note that one of the Gray's pastures, 44,000 acres of a relict Pleistocene lake bed that stretches into Mexico, would hold three Manhattans. But who could even think of such an obscenity?

To experience this place is to realize that we profane it by trying to quantify its magnificence in terms of how many endangered plants and animals have been counted within its boundaries. Those who know the Gray know there is more to it than the sum of its parts—more to it than Sanborn's long-nosed bats, night-blooming cereus, and endangered sand-dropseed grassland communities. There is the pure, mind-altering sweep of uninterrupted open space. To look across the Gray is to feel the exhilaration of flying, even though you're standing on the ground. Seeing a rock at one prominent vantage point that has been worn smooth, it is easy to imagine that even the ancients were awed by this breathtaking expanse and that, over the centuries, they too paused to sit and look. Big as it is, though, this landscape is still vulnerable. "I've seen places like the Gray disappear under concrete," cautions Gary Bell, who came to the ranch from a Nature Conservancy preserve in California that had concrete and condos crimping its toes.

The threat to the Gray that the Nature Conservancy was trying to avert by buying it was not subdivision, at least not yet. At 200 miles from Tucson and El Paso, the ranch is too far from any urban centers to end up as a grid of tract houses or ranchettes. The biggest threat to the Gray, and to most ranches of comparable size, is fragmentation—the breaking up of a large land holding until it loses its coherence and cohesiveness; until it becomes too split up to function as a whole, as an ecosystem. Few people have the wherewithal (and few corporations have the incentive) to buy a chunk of land this big and remote. If they do, it's generally on impulse—because it's there. But impulses like that rarely last. Since the ranch had been acquired by its Mexican billionaire owner as collateral on a debt, it was logical to assume that he would do whatever was necessary to recover his investment. With no particular reason to keep the Gray in one piece, each day dawned as the day that could end the ranch's existence as a political as well as an ecological unit. When the Conservancy learned that the ranch's owner would consider an offer, they jumped.

The Conservancy entered into serious negotiations for the ranch, and all of those interested in the fate of the Gray were convinced that if the Conservancy did manage to acquire it, it would turn around and sell the ranch to a government agency as soon as one had the money to buy it. New Mexico Senator Jeff Bingaman affirmed as much by requesting that Congress appropriate money for that purchase while the Conservancy was still negotiating. Such a turnaround would have come as no surprise—the Nature Conservancy has done the same with many of its other large acquisitions. One such place, the San Pedro Riparian National Conservation Area, is a mere 100 miles west of the Gray in Arizona—close enough for the rural peoples who live near the Gray to hear their Arizona neighbors howl about increased federal encroachment. When the Conservancy finally managed to purchase the ranch in January 1990, the surrounding community was all ears.

In spite of all the commitment and optimism the Conservancy expressed by buying the Gray,

remaining its sole owner was not a realistic option. The price tag finally agreed upon was eighteen million dollars, and though the Nature Conservancy is the richest of all environmental organizations, it needed someone to help it maintain that liability. With regard to who that might be, "All possibilities were on the table," says John Cook, a Conservancy vice president appointed by John Sawhill as project director for the Gray Ranch Project. "Pretty much every agency was interested in the Gray."

In New Mexico, the government owns 43 percent of the land. That's more than enough of a government presence for the rural people of that state, who nurture a love of independence and privacy, even if some of them do take advantage of cheap federal grazing fees and subsidized water. Lines of resistance to the hand of government have been in place among these mountains and prairies since the Spanish came here four centuries ago. One of the families that lives near the Gray has claimed ownership of its land, including all things on it living or dead, from the center of the earth all the way to the heavens. In Catron County, to the north, an ordinance was passed that makes any attempt by a federal employee to cut the number of cattle that a rancher can run on *public* land a violation of that rancher's civil rights.

When the Gray's neighbors got wind of the fact that one of the alternatives being considered by the Conservancy (a U.S. Fish and Wildlife preserve) would bring an estimated 65,000 recreationists into their rural enclave, their opposition to the project, and to the Conservancy as well, went ballistic. To them that meant 65,000 strange faces driving blankly by, probing their driveways and ranch roads, climbing over their fences on the assumption that all the land was public. And, worse yet, it meant 65,000 politically active urbanites saying "You mean they let them graze cattle on this?" The Conservancy might as well have announced that they actually *were* going to plop down three Manhattans in the Animas Valley and throw in half of Rhode Island to boot.

To add insult to prospective injury, many of the options being discussed for the Gray would end cattle ranching there. To the people of southwestern New Mexico, that was standing the truth on its head. To them, the Gray is as magnificent as it is because their ranching culture has kept it that way. The only reason anything was left to risk, they would tell anyone who took time to

" ...it is easy to imagine that even the ancients were awed by this breathtaking expanse and that, over the centuries, they too paused to sit and look."

listen, was because the ultimate absentee land-owner, the U.S. government, had not yet had a chance to mess it up.

An intense lobbying campaign was sufficient to block Senator Bingaman's request for federal funds, so, with the possibility of any other federal buyout becoming more unlikely with each irate phone call and bristling letter, Ted Turner and Jane Fonda entered the picture. Turner has become a major holder of western rangelands through an aggressive campaign of acquiring the biggest and best of them, which he then stocks with bison rather than cattle. Since the Gray was a natural for Turner's growing stable of ranches, he offered himself as an alternative to federal owner-ship and entered into a marathon of negotiations over the terms of the deal. The Gray's neighbors were no more enamored of Ted and Jane than they were of federal bureaucrats, however, and their howls of opposition intensified. After Turner bowed out to buy a different ranch in central New Mexico, the Conservancy decided the time had come to put the effort to sell the Gray on a back burner until the community issues could be resolved.

Making that decision came easily, according to John Cook, because the Nature Conservancy was going through some changes of its own while all this was happening. "We were beginning to realize that park-preserve-style efforts aren't going to be enough to close the gap on the care of areas worthy of protection over the next 100 years," said Cook. While the Conservancy was thinking the matter over, it stood fast and kept its cards close to its chest. As they thought, their point of view continued to evolve. Finally, they made a momentous decision.

"We came to think that this was the place to see if private ownership could be equal or possibly even superior to public ownership in achieving both conservation goals and rural economic goals," declared Cook.

What happened next, according to Cook, was not a matter of being forced by local pressure to take second best. "In Washington there's so much focus on public lands management, but when you get down here you find that the land is really a seamless mosaic of public and private ownership." Cook affirmed that the Conservancy's board of directors decided that preserving bio-diversity and functioning ecosystems can't be achieved by focusing on the public part of that

mosaic alone. "For one thing, over 50 percent of listed [threatened and endangered] species are on private land," he noted. Realizing that, he added, is why the Conservancy came up with the Last Great Places initiative in the first place.

"In my view," wrote Conservancy president John Sawhill, in kicking off that initiative in the Nature Conservancy's magazine in June of 1991, "'Last Great Places' must involve local citizens to be successful. No one knows these places better nor has a more direct interest in them.... No word better typifies 'Last Great Places' than partnership."

What was needed, then, was a buyer who not only had the respect and support of the rural community but who shared the same concerns and dreams as the Conservancy. After Ted Turner walked away, the Conservancy got a call from Drum Hadley.

The community had turned to Hadley during the heat of the controversy to see if he could somehow find a way to keep the Gray from becoming a foreign beachhead within their island of isolation. The owner of a ranch in nearby Guadalupe Canyon, Hadley was one of them—but he was one of them with connections. He had spurned those connections but not cut them when, as a young man thirty years ago, he turned his back on a future in the family business of Anheuser Busch, and, climbing through the seven strands of barbed wire that mark the international boundary with Mexico, headed south of the border with a head full of poems, a saddle, and a bedroll to look for a job punching cattle. A friend of Gary Snyder and other spiritual seekers of the sixties, Hadley was searching for a way of living with the land that grew from it rather than was imposed on it. He found that "way" in Mexico, he told writer Alan Weisman in an interview for the book *La Frontera: The United States Border With Mexico,* "because Mexico provides the con-tinuum of the traditions created in the image of the earth...a consciousness of the land, weather, and animals. This doesn't mean any kind of eco-logical consciousness, but a consciousness of what their value imposes on human beings."

When Hadley came back across the fence, he brought that knowledge with him. Realizing that the knowledge would wither without a piece of land to nurture it, he bought a small, isolated ranch in Guadalupe Canyon, one step north of the border and one mountain range west of the Gray, then he began to make his management of

"…he moved the fences that defined the ranch's pastures so they conformed to natural boundaries— ridgelines and watershed divides…"

that land a reflection of its own natural patterns. First, he moved cattle out of the riparian area along the canyon bottom that had been grazed to near sterility over the years. Next, he moved the fences that defined the ranch's pastures so they conformed to natural boundaries—ridgelines and watershed divides—rather than the imaginary administrative lines defining sections and townships that they normally follow.

Ridgeline fences made it easier for cowboys to gather cattle, doing away with over-the-hill blind spots. This was as much a matter of spirituality as utility for Hadley. When it comes to the land, as simple a matter as changing the location of a fence holds deep significance for him. "You work with the land rather than against it," he mused one day, as we drove up Guadalupe Canyon to visit the house he had built there. "You don't impose quadrilinear metes and bounds, concepts developed in Europe, on the rough-canyon, mountainous landscape of the West."

In addition to restructuring the role of humans in relation to the land, Hadley has also become interested in restoring one of the natural forces that humans have banished from it—wildfire. A man Hadley met at Guadalupe, and came to know and love, made him aware of the changes that the suppression of natural fire has brought to the ecosystems of the West. Hadley refers to a horseback ride with this teacher to enjoy the view at a high overlook as a watershed in his awareness of the crisis facing the West's open spaces. He told me about that ride as we stood at the same overlook with the sun setting over our shoulders, bathing the canyon and the mountains of Mexico beyond it in a golden glow.

"My old friend Walter Ramsey, who was raised in Guadalupe Canyon and now would be over ninety, brought me here when I first moved to the canyon and said, 'Y'know, Drum, you used to be able to count the cedar trees you could see from here. Now there are thousands beyond counting.' Then we rode on a little farther and came to a stockwater tank in a flat that was covered with mesquite and had erosion ditches fifteen to twenty feet deep, and Walter said, 'Y'know, Drum, this used to be a pretty little grama grass flat.' That's when I began to wonder about the future of ranching in the West and about the future of its open spaces, too."

Ranchers have been fighting a war against this "invasion" of woody species—cedar (juniper),

mesquite, sagebrush, rabbitbrush, burroweed—virtually since ranching arrived in the West. In areas where these small trees, shrubs, and bushes achieve maximum density given the available moisture, the grass frequently disappears, erosion starts, and gullies begin to form. Ranchers and old-time range managers blame this change from grass to wood for everything from the lack of water in western streams (deep-rooted trees suck it out of the ground and transpire it into the air) to the lack of money in their bank accounts. Historic photographs show a massive increase of these plants in many areas of the West over the last century. The verdict of those who try to explain this increase is virtually unanimous: In part, if not in its entirety, they say, it is due to the banishment of fire from the landscape by the herds of cattle and sheep that consumed the natural fuel that nourished those fires.

Unwilling as much as unable to address this cause, but determined not to be pushed off the land, ranchers began their counterattack in the 1940s. To get the job done on a scale that made it economically worthwhile, they reached for heavy artillery. One of the weapons of choice was a brace of D-9 bulldozers growling across the dusty landscape with a huge ship's anchor chain stretched between them that clanked and scraped and jerked the land clean of woody interlopers. Herbicides dropped from planes, controlled burns, and hordes of firewood cutters made up the rest of the arsenal. Since the 1940s the BLM estimates that more than six million acres of the western rangelands it manages have been subjected to this "treatment," although that figure includes a significant amount of repetition. Add to that the U.S. Forest Service and private acres that have been thus treated and the result is a checkerboard of alternately cleared and uncleared sections that pollutes the view from just about any rangeland overlook in the West. A closer view of these "improvements" reveals a landscape littered with the carcasses of uprooted trees, with a new generation soon to be ready for bulldozers and anchor chains growing through the woody skeletons of their predecessors.

"I'll be dead before this whole damned place turns into one big brush pile," one fatalistic warrior remarked.

Recently, as the political environment has become increasingly hostile to such sweeping intervention, some ranchers have adopted methods

"You don't impose quadrilinear metes and bounds, concepts developed in Europe, on the rough-canyon, mountainous landscape of the West."

Cattle graze among native grasses, oaks, mesquite, and yucca on the wide open spaces of the Gray Ranch.

that are less destructive and yield less artificial results. On his K-4 Ranch in central Arizona, John Kieckhefer has traded metal-tracked bulldozers for rubber-tired tractors that roll around the landscape popping out trees whose smaller size labels them as encroachers. Kieckhefer's drivers are instructed to leave the older, larger resident trees to restore the natural, preinvasion mosaic. Behind them, the native grasses resprout without the help of seed drills or broadcasters.

Kieckhefer has tried to use the more natural force of fire to keep his grasslands open, but air-pollution laws and liability concerns that arose because of his ranch's proximity to an area of extensive suburbs made that an impossibility. Instead of being left to burn, fires on the K-4 are quickly suppressed. Ranchers like Hadley, on the other hand, with fewer suburbs to contend with and with their eyes set on restoring natural processes, are moving to reintroduce wildfire to maintain a balance between trees and grass.

Before European society came to the West, fire was one of the dominant forces that shaped the region's grasslands. Most grasses have their tissue-generating mass at or slightly below ground level, where it has evolved over millions of years of coexisting with fire and coevolving with grazers. Because of that protective physiology, those grasses survive all but the hottest of flames. Woody plants, on the other hand, have growth tissue exposed above ground in limbs and stems. They suffer harm and are even killed by fire. The frequency of lightning strikes is high all over the West, and especially so in the Animas country that is home to the Gray. That, in itself, has made the grasslands of the ranch naturally inhospitable to woody species; but anthropologists and ecologists now tell us fire may have visited the Animas even more frequently than lightning could account for.

Spanish explorer Cabeza de Vaca, in recording his travels in southwestern Texas in 1528, wrote, "[The Indians] go about with a firebrand, setting fire to the plains and timber." They set fires, Cabeza de Vaca observed, for a variety of reasons—from driving off mosquitoes to forcing small animals, even deer, into the open, where they could be killed for food. The Indians also used fire, he recorded, to deprive the deer of forage in one area, so they would be compelled to go where the Indians wanted them to. Other writers tell of Indians using fire to burn stagnant grass and to create a flush of new growth to attract game

or, after the Spanish brought horses, to feed their ponies.

Ray Turner, retired co-director of the Desert Lab at the University of Arizona in Tucson, has dedicated his life to studying the changes that are occurring in the ecosystems of the Southwest, using the technique of repeat photography to open a window in time through which we can all glimpse those changes. In a pioneering piece of ecological detective work entitled *The Changing Mile*, published in 1967, Turner presented ninety-seven photographs, most of which were taken around the turn of the century, and displayed beside them repeat photographs of the same locations taken in the early 1960s. Many of those photographs chronicle the same loss of grasslands and march of woody species pointed out by Walter Ramsey. Since *The Changing Mile*, Turner has continued to seek out old photographs of the Southwest and rephotograph the same locations to give us a time-lapse portrait of the changes that have occurred within the Southwest—at least since cameras have arrived.

Turner says that, before fire was artificially suppressed here, the grasslands burned at least once or twice every ten years. When the Spanish, and later the Americans, came to the region, they brought domestic livestock with them, and where those animals consumed the land's natural fuel load they effectively banished fire. Fires that managed to start in spite of that were put out to "save the range." The result, says Turner, was just the opposite. Fire suppression encouraged the growth of woody species at the expense of resident grasses. In some places, this conversion was so complete that grasslands were replaced by monocultures of trees devoid of an understory and capable of supporting only a narrow band of diversity. These juniper and mesquite barrens then became ecological wastelands.

Spurred by the concern raised by Walter Ramsey and the knowledge contributed by Turner, his father-in-law, Hadley joined with twenty-five local ranchers to start the wheels turning toward reintroducing fire into the still-remote ecosystems of southwestern New Mexico and southeastern Arizona. First, the ranchers agreed to observe a let-burn policy toward any natural wildfires that occurred on their lands. Second, they moved to bring in federal and state land managers to help find a way through the maze of conflicting and overlapping regulations and jurisdictions that

"Most grasses have their tissue-generating mass at or slightly below ground level, where it has evolved over millions of years of coexisting with fire and coevolving with grazers."

make large-area prescribed burns virtually impossible. The result of that coming-together was the Malpai Borderlands Group (MBG), an association of ranchers, agency land managers, and urbanites united in their common concern for the future of open lands in the West. The formation of the MBG, in effect, created an unofficial planning area of more than a million acres of public and private land in southwestern New Mexico and southeastern Arizona.

While the MBG was coming together and forming its policy toward wildfire, the Nature Conservancy was doing some serious thinking of its own about wildfire and its place on the Gray. In an article for the group's newsletter, writer Ben Brown mentioned that the Gray's grasslands may be as pure as they are because the ranch was never grazed hard enough to exclude fire; nor were fires suppressed when they did start there. While it owned the ranch, the Conservancy also studied the natural fires that occurred there with the thought that, someday, human-caused fire might have to be reintroduced to maintain the integrity of its grassland ecosystems. The fact that Drum Hadley, the community's choice as owner of the Gray, was thinking the same thing helped bring the two together. But it was the riparian area at his Guadalupe Canyon Ranch that clinched the deal.

"I don't know a damn thing about cattle ranching, but when I took a ride down this canyon, I knew that Drum was the man to do the Gray," stated John Cook, as he nodded toward the white-gray-barked Arizona sycamores and silken tufts of deer grass along Guadalupe Creek. The canyon had blossomed under Hadley's management, and it was obvious to anyone who bothered to look.

Conservancy President John Sawhill came to Guadalupe Canyon, too, and was also convinced that Hadley was the man for the job. All that stood in the way was the small matter of several million dollars needed to complete the purchase. Hadley's son, Seth, helped take care of that obstacle by suggesting that he and his father pool their inheritances to form a foundation that could then attract enough additional support to buy the Gray and manage it. With the help of Drum's mother, Puddie Hadley, the brainchild of that bit of creative problem solving, the Animas Foundation, was born. This private, nonprofit foundation also solved another problem. It made it possible to

attract the expertise needed to give the Gray the quality management necessary to take care of its extensive natural treasure chest.

When the Animas Foundation was first conceived, Ray Turner thought enough of the idea to offer to serve on its board of trustees. John Cook asked for, and was granted, a year's sabbatical from his job at the Washington headquarters of the Nature Conservancy to help the fledgling foundation take over the reins of the Gray. "I'm here as toolmaker…for the long haul," declares Cook. Such a collection of expertise, vision, and commitment made it possible for the foundation to enter into a partnership with the University of Arizona to jointly fund a senior research position at the university that will eventually become an endowed chair. The foundation also enabled the ranch to install a state-of-the-art computer mapping system, known as the Geographic Information System (GIS), at the ranch, which will also be used on other lands affected by the Malpai Borderlands Group, as well as on the Gray.

"In the end," says John Cook, "we didn't sell the Gray to a cattle rancher, we sold it to an institution with a charter, a board of trustees, and nonprofit status—an institution that has agreed to own the ranch forever, to never let it be developed, and to continue a very extensive monitoring program in perpetuity to ensure that the Conservancy's goals are being met."

But the creation of the Animas Foundation wasn't the only change this piece of creative problem solving brought to the Borderlands community. It changed the way a significant part of that community relates to the land.

Before the transfer of the Gray Ranch to one of their own, the Malpai Group had been a back-porch gathering of local ranchers working together to achieve nuts-and-bolts goals like bringing telephone service to the Borderlands, expanding school bus routes, and bringing in rural electricity. When they did get together at Warner and Wendy Glenn's Malpai Ranch to work on these initiatives, they would take the opportunity to grumble about issues that came from the "outside"—the push to increase grazing fees, the movement to declare some of their land wilderness when they felt it was more wild without designation, the laws that kept them from allowing fires to burn as they once had through these magnificent grasslands. Occasionally, they would send off a letter about these issues, too;

"…the Gray's grasslands may be as pure as they are because the ranch was never grazed hard enough to exclude fire; nor were fires suppressed when they did start…"

Green grass thrives while a mesquite tree has succumbed in the aftermath of natural fire on the Gray Ranch. In 1993 and 1994, nearly 100,000 acres of rangeland burned in lightning-started fires on the Gray and other near-by ranches managed by members of the Malpais Borderlands Group.

but they did it individually, not as Drum Hadley urged them—as an organized group with a letter-head, a plan of action, and a growing constituency.

The formation of the Animas Foundation changed all that. It created the very type of broad-based working group that the community had, up till then, barely dared to dream. And it brought new players onto the scene—scientists, urban environmentalists, government bureaucrats. Most importantly, it brought them here in an atmosphere of respect and commonality that was unique in the embattled rural West. The results were what they usually are when people who consider themselves adversaries find out that they really can get along and work together: It created an atmosphere of optimism and enthusiasm that was infectious, almost giddy. "We started feeling good about the future," Wendy Glenn said in typical understatement.

The Malpai Borderlands Group set up a charter and applied for nonprofit status in 1993, three years after the Animas Foundation was formed. The expanded area affected by the MBG, along with the 320,000 acres of the Gray, totaled nearly a million acres that its members either owned or were managing as grazing allotments. After the formation of the MBG, other more land-based initiatives came quickly.

The first of these was the untying of the knot of red tape that had obstructed the return

of fire to the Gray Ranch and to the rest of the Borderlands ecosystem. The authority to approve the prescribed burning program the MBG put together rested with nine government entities: two BLM Districts, the U.S. Forest Service, the state lands departments and wildlife departments of Arizona and New Mexico, and the U.S. Fish and Wildlife Service. In addition, the program had to be coordinated internationally with Mexico. Since most of the United States entities had signed on with the MBG as cooperators, and since Mexico was now watching this process with increasing interest, what was once impossible was now almost easy. "We got the job done in eight months," said a member of the MBG board. The first prescribed burn was scheduled to take place in May 1995. In the meantime, lightning-started fires left to burn naturally revitalized nearly 123,000 acres of the Gray and other ranches of the Borderlands area in 1993 and 1994.

Another innovative program put together by the MBG and the Animas Foundation is based on a practice traditional in rural communities—the practice of neighbors who have sharing with those who have not. In this case, however, the program (called a "Grass Bank") goes beyond pure neigh-borliness to address more contemporary concerns.

Ranchers whose land is in need of rest from grazing for any of a number of reasons often find themselves faced with alternatives that are unattractive or ineffective. If they want to rest the land to restore its vitality, to avoid ecosystem damage caused by grazing during drought, or even to accumulate enough biomass to carry a fire, their choices are to lease other land and stretch thin margins even thinner, cut their herd and lose money, or continue to pound the land and hope the rains come before permanent damage is done. The Grass Bank gives ranchers in this situation the option to trade a conservation easement on their ranch for temporary grazing on another cooperating ranch that has enough forage for both herds. The end result is win–win–win: The rancher benefits by having his land rested and restored; the rural community benefits by having more land protected from fragmentation; and the natural community benefits on both counts. The Gray Ranch with its extensive reserves of grass and light stocking rate stands to play an important role in this program.

While most ranchers cringe at the idea of finding an endangered species on their ranch, at least some members of the MBG have quite a different view of the matter. "If someone finds an endangered species that lives on my ranch and nowhere else, I figure that means I'm doing a good job of managing my land," says MBG rancher Warner Glenn. Another rancher from the Borderlands area has been hauling water to a population of threatened Chiricahuan leopard frogs living in his wet-weather stock ponds in an effort to sustain them through the area's frequent dry spells. Populations of these rare frogs living in natural ponds are preyed upon by introduced bullfrogs, so the artificial impoundments have become critical to their survival. The MBG is planning a way to provide permanent water to those ponds and to help the rancher improve his overall land management practices to stabilize the frogs' habitat.

If you're thinking that these are the kinds of programs one would expect to find on a nature preserve rather than on a collection of cattle ranches, consider that a nature preserve is exactly what the members of the Malpai Borderlands Group believe they are creating. "We call it a working wilderness," declares Bill McDonald, a rancher who is a member of both the Animas Foundation's and the MBG's board. McDonald says it's a matter of both sides working to achieve goals that are no longer "ours" or "theirs," but that have become goals for the entire community.

"…it brought new players onto the scene—scientists, urban environmentalists, government bureaucrats. Most importantly, it brought them here in an atmosphere of respect and commonality…"

The result is a situation as liberating as it sounds; where it's okay for a rancher to save threatened frogs or for an environmentalist from back East to work to sustain cattle ranching.

The players in this scenario—the members of the Borderlands community, the Nature Conservancy, and a number of government land managers—have worked hard to keep the debate over the future of the Gray Ranch and their homelands from deteriorating into a war of opposites, as so many western lands debates do. As a result, all who have kept their cards on the table here have come out winners. By not insisting on turning the Gray into a 320,000-acre wildlife preserve run by the government or some other outside entity, the Nature Conservancy has helped to create a "working wilderness" three times as large. And this preserve is growing, with more ranchers wanting to join.

By reacting with offers of constructive collaboration rather than doctrinaire opposition to the entry of a national environmental group as a major player in the Borderlands, the region's ranchers have succeeded in remaining at the table where the future of the lands they call home will be charted.

Most important of all, by working together to create this outstanding example of community building, Drum Hadley, the Nature Conservancy, the Animas Foundation, and the MBG have shown the rest of us a way we can have rangelands that are ecologically, economically, and socially sustainable without relying on the government to do it for us. That is no small achievement.

"If we try to imagine the kind of thinking which would lead people…to work for the "higher common good" we see them acknowledging to one another that they all want to live well on a certain, very definite part of the earth."

DANIEL KEMMIS
from Community and the Politics of Place

ORME RANCH, ARIZONA

Left to right: Ben Powers, Donn Rawlings, Jack Turner, Emerson "Casey" Jones, J. B. Kessler, Diana Kessler, Dwayne Warrick, Alan Kessler, Jigger Kessler, Kit Metzger, Doug MacPhee, Patrick Boles, Mary Ellen Hale, Don Charles.

THE
⅃ RANCH

and Diana Kessler

ARIZONA

On Trust, Teamwork, and Tobosa Grass,
Coyotes and Antelope, Goals, and Holism

T HE ORME RANCH ENCOMPASSES an area of rocky malpais slopes and Hush Puppy tan tobosa flats about halfway in distance and elevation between Arizona's gold rush Bradshaw Mountains, south of Prescott, and the Verde River canyon that drops toward Phoenix. Someone once said that tobosa is as much a bush as it is a grass. When it's dry it resembles the stiff, curly, woody stuff that dishes are sometimes packed in. It's probably just as hard to eat too, except for the few days out of the year that it's green and growing. Faced with a forage source like that, Alan Kessler says what he needs to graze the Orme the way it ought to be grazed is about 3,000 head of cattle two weeks out of the year—and none the rest of the time.

Kessler is tall and blond, with a build that's an obvious indicator of hard work. With his hair parted in the middle and a bushy mustache that curls a bit on the ends, he looks like he'd be right at home in a barbershop quartet. Kessler and his wife, Diana, who is poised, perceptive, and involved in ranch work as well as family, took on the job of running the Orme in the early 1980s. At that time the Prescott National Forest, on which the ranch has a 20,000-acre grazing allotment (in addition to its 4,800 acres of state lease land and 1,200 acres of private land), was in the process of shepherding a forestwide, ten-year management plan through the maze of appeals brought by a variety of interest groups. The task was proving to be more than a little rocky, as it was for many national forests at the time, and it was sprouting controversies like weeds in a bulldozer track. As the process dragged on, members of environmental groups were

The Orme Ranch Management Team monitors an area on which the grass was once so sparse cattle stayed on it only as long as it took them to walk across it. "We can get about a six-day graze out of it now," declares Alan Kessler.

becoming increasingly critical of grazing on public lands, and that, in turn, was jabbing at the Orme's panic buttons.

With the level of controversy ratcheting steadily higher, Kessler decided that the best way to lower the heat, at least with regard to his ranch, was to get some of his critics to come and see that he wasn't doing such a bad job after all. That was the best way he knew of getting the situation under control. "We wanted the people who were opposed to at least understand what we were doing," he recalls. With that in mind, he approached local chapters of the Sierra Club, the Audubon Society,

and the Arizona Wildlife Federation to see if some of their members were interested in coming out to the ranch for a look-see.

Kessler's outreach program had limited success until he met Donn Rawlings, an instructor of English Literature at Yavapai Community College in Prescott. At the time, Rawlings was vice president of the Prescott Audubon group and had attended several of the forest plan appeals meetings as a member of both Audubon and a local group, the Prescott National Forest Friends. Alan quickly developed a strong respect for Donn because of the way he worked with people and because Alan perceived a deep sense of integrity in the man. So, he invited him not only to come to the ranch but to be part of its management team. "I had a feeling that Donn was really interested, and that he wasn't just going to sit there and take shots at us," Alan remembers. He was pleased when Rawlings took him up on the invitation.

The Orme Ranch Strategic Team that Alan asked Donn to join had once been part of the forest service planning process, too. It had started out as a working group and an interdisciplinary team of agency experts (ID Team) that the forest service had put together to hammer out an allotment management plan (AMP) for the Orme's forest service grazing lease. Federal regulations require that each public lands grazing allotment has an allotment management plan, put together with full public input and renewed every ten years. Many are long overdue. The AMP for the Orme Ranch was completed in 1985.

"After we got our plan approved, we decided to keep on meeting, though we weren't required to," Kessler told me as we sat in his simple living room with *Audubon* magazines mixed with Arizona Cattlegrowers newsletters and cowboy songbooks on the coffee table. "I guess we were just used to each other by then." Without the government calling the shots, most teams break up and never look back. This one didn't break

up, but it did change. One of the first changes that Alan noticed was that, after the team members started getting together on their own, the atmosphere of the meetings became more friendly and less formal. Instead of exchanging stares, as if they had been plotting how they were going to slam dunk one another since the last "public input session," greetings were warm and personal, even among people who felt quite differently about the issues. Instead of shaking hands, some began exchanging hugs. They made more small talk, about their personal lives, about the grand-kids. "Afterwards we'd all just hang around out in the yard and visit," Alan says.

But it takes more than friendly conversation to keep a group like this coming back every two months for ten years. Most of those who did keep coming were agency people, but a few other members of the community started showing up, and then there were relatives and other ranchers. All are busy people, and team meetings are held on weekends, a time most of us don't give up without a fight. While it seems trite to say they kept coming back because they wanted to, no one forced them. They weren't getting any recognition for it. Some were even being criticized by their peers for talking with the "opposition." When I asked a few of the team members why they were willing to give up their weekends to

"...they were doing something special here, something outside the bureaucratic process, something more important even than having fun."

Tall tobosa grass makes good hiding cover in which antelope can bear their fawns out of sight of hungry coyotes. At their management team's behest, the Orme Ranch has committed to building fences and moving cattle in such a way that leaves plenty of tall grass for antelope and their fawns.

come to a meeting to talk about the same things they talked about at work, they told me it was because they believed they were doing something special here, something outside the bureaucratic process, something more important even than having fun.

Kyle Cooper works for the Arizona Game and Fish Department. He comes to the meetings, he says, because he's pleased to work with a rancher who is interested in more than just cattle, and because he thinks the team can help further his goal of helping the antelope population of Wildlife Management Unit 21 recover to the unit's carrying capacity. Because of Cooper's input, Kessler has agreed to build some new fences and to alter his grazing rotation while the antelope are fawning. As a result, the does will have a greater range of choices of where to bear their fawns—in tall grass to hide them from coyotes, or among the cattle, whose presence in large numbers also serves as a deterrent to predators.

Carla Staub, a University of Arizona graduate student in a cooperative work-study program with the Prescott National Forest, is young and bright, full of energy and enthusiasm. "It's spelled with a C," she corrects, as she watches over my shoulder while I take notes. Carla says she has changed a lot within the last year, during which she attended a personal growth seminar on adopting a win-win approach to life and put in a weekend at a workshop on Holistic Resource Management. Before that, she says, she was working for a wildlife agency and was opposed to grazing livestock on public lands in a way common to people in wildlife positions. "I was isolated and, talking to people who felt the way I did, I got negative and stayed that way," is how she put it. After the two seminars, Carla's opinion changed, and she doesn't hesitate to say she's glad it did. "I'm not so negative now about cows and ranchers," she said, "and I feel a lot better about it."

Donn Rawlings said his motive for joining the team was to learn how a progressive ranch managed its public lands, and, as a lover of nature and remote places, to become better acquainted with the public and private land that it encompassed. Though Rawlings grew up on a small ranch in northern Montana, he is no rancher "wannabe" and makes that clear without hesitation. For him, the main issue here is deciding which lands are suitable for grazing and which are not. Rawlings feels lands are unsuitable where the

ecosystem is too fragile or too valuable as wildlife habitat to be required to support cattle. "There are places that have been destroyed by grazing, and I can show you some," he states. "I think those places need to be declared off-limits to livestock."

Other members of the team share Rawlings's opinion: that livestock grazing can be done right and it can be done wrong, and that there are places where it shouldn't be done at all. Within that sharing there is room for a lot of diversity.

When I asked Jack Turner, a hydrologist for the Prescott National Forest, why he has been attending these monthly meetings for three years, he said he does it because he believes that the process of collaboration and teamwork offers more promise for solving environmental problems than the old "squeaking wheel gets the grease" approach. As an employee of a federal lands agency at a time when controversy over land management decisions has him and his agency all but at a standstill, Turner has plenty of experience on which to base that judgement.

For Jack, just keeping these meetings going, keeping the lines of communication open between people who otherwise wouldn't talk to one another, and helping them discover how much common ground they share are all important enough for him to give up a weekend every month or two. They're important enough that he has been willing to invest the time even if the payoff hasn't come immediately—just as long as he feels that the team is growing in strength and that its members are becoming more open and more willing to work together.

Mary Ellen Hale is a massage therapist in Prescott. When she joined the team she was a student at Prescott College, an innovative private school with an extensive environmental education program. Mary Ellen's feelings for the land are deeply spiritual. They focus on holism and healing. As part of her course work she attended a Holistic Resource Management class at the Orme Ranch, where she got to know the Kesslers and learned about their management team. After the course, she decided to become a member of the team because she felt that Allan and Diana shared her feelings for the land and that the team was a manifestation of those feelings.

Mary Ellen is also deeply committed to non-confrontation. Though she is a committed environmentalist, when environmental groups send

"…keeping the lines of communication open between people who otherwise wouldn't talk to one another, and helping them discover how much common ground they share…"

her solicitations that are phrased in terms of good versus evil and us against them, she underlines the confrontational words and sends them back—without any contributions, but with a note that says, "You say these things and expect me to believe you're looking for solutions?"

Mary Ellen is clear about which approach she believes offers the best hope for finding solutions to environmental problems. "The team model is a cooperative one," she explains. "The other model [the confrontational one] is based on competition for competition's sake, not for the benefit of the land or the environment."

Many of us in today's atmosphere of high pressure, high risk, and constant crisis consider the approach Jack Turner and Mary Ellen bring to environmental issues as lightweight and wimpy—a sort of appeasement that makes the world worse instead of better. To us, today, the stakes appear so high that to accept anything less than victory is to lose. To compromise is to sell out. What matters is results, and the only result that matters is winning. When policy decisions are made about environmental issues, each side assesses them as a victory or a defeat and, based on that assessment, makes plans either to retaliate or prepare for a counterattack. We pick our leaders, give our money, and voice our praise on the basis of who is best at defeating the opposition.

But Mary Ellen Hale, Jack Turner, and some of the other members of the Orme Team will tell you that they have found that, in order for them to win, it isn't necessary for someone else to lose, or even to compromise—their team and what it has achieved is living proof of this concept.

On the Orme Ranch, winning is measured in terms of goals. Goals are defined in terms of the land—in terms of its health, vitality, and diversity—not in terms of what we ought to be doing on it. Team members there have found that win/lose situations arise all too easily when they address land management decisions from the "let's decide what to do" approach. If the team decides to do what some of its members want, almost invariably it means that some of the others don't get to do what they want to do. One side's winning then becomes inextricably attached to the other side's losing, and they're in conflict before they get started. If, on the other hand, the team sets goals that are defined in terms of the condition of the land, such as having more healthy riparian areas or more biodiversity, they've found

"If we work toward a goal that is defined in terms of the land and not in terms of which one of us gets our way, it becomes much less necessary for either of us to lose or even to have to compromise."

that it's quite possible for everyone to get what they want without anyone coming out a loser. Think back to those students at the Verde Valley School and that photograph of Phil Knight's stretch of Date Creek.

If we work toward a goal that is defined in terms of the land and not in terms of which one of us gets our way, it becomes much less necessary for either of us to lose or even to have to compromise—to be pragmatic maybe (no raising bananas in Minnesota), but not to compromise.

One of the first things the Orme Ranch's strategic team did when it moved beyond the forest service planning process and began meeting on its own was to put together a list of goals defined in terms of the land. While team members were contributing to the list, nothing was off-limits. No one was asked to pull any punches or to water down what they asked for. Each was encouraged to say just what they wanted, and when they did it was written down.

Since that initial session, there have been a few changes. But as it exists now, the list of goals for the Orme Ranch includes seventeen entries placed under three categories: quality of life, production, and landscape.

Quality-of-life goals include "To improve and sustain the health, diversity, and productivity of the land" and "To have responsible community participation in developing a healthy and diverse ecosystem."

Production goals include "To produce diverse and abundant plant and animal communities" and "To produce and maintain healthy riparian areas and flowing streams."

Landscape goals include "To have a healthy grassland with many species of warm and cool season grasses, browse, and forbs" and "To have abundant animal life, i.e., antelope, deer, small mammals, predators, birds, reptiles, and insects."

At their most recent meeting, team members rated the success the ranch has had in achieving their stated goals. On a scale of one (best) to five (worst), improving the health, diversity, and productivity of the land received the highest rating (1.8). Making a profit got the lowest success rating of all (4.3). Across the board, success in achieving landscape goals was rated significantly higher than both production goals and quality-of-life goals. In other words, the Kesslers have been more successful at improving the land than they have been at enhancing their own quality of life and the quality of life of the ranch's owners. That stands in stark contrast to the caricature of western public lands ranchers as a fat cat dripping dollars and getting rich at the taxpayer's expense. As welfare ranchers, the Kesslers are a definite flop.

"...the team has been a success, not just because the ecosystem has improved steadily over the time it has been directing the management of the ranch, but because that improvement has been achieved at bargain prices to the government."

Dwayne Warrick was the top range and wildlife staff officer for the Prescott National Forest for thirteen years, until he retired in early 1994. He is still a member of the Orme Team and has been since its inception as a forest service public input group ten years ago. Warrick says he feels the team has been a success, not just because the ecosystem has improved steadily over the time it has been directing the management of the ranch, but because that improvement has been achieved at bargain prices to the government. "They've done a lot on their own," Warrick says of the Kesslers and the ranch, "and the agency has only spent $4,500. A lot of ranch development plans run in the hundreds of thousands of dollars, even without all the changes the Kesslers have made."

When Alan Kessler came to the Orme in 1981, the Prescott National Forest had recently completed a study of the ranch's carrying capacity. That study showed that the Orme's forest service allotment was overstocked by 40 percent. "They had been chasing green grass for years," remarks Doug McPhee, Prescott National Forest range conservationist. "The result was that they had done significant damage in areas that green up the earliest and had been underutilizing the rest of the ranch." McPhee recommended an interim recovery program, to which the ranch agreed. Kessler came on as manager in the middle of that program and carried it to completion. Then he continued improving the ranch to the point where its carrying capacity in 1993 was more than twice what it

was when the forest service recommended cutting it in 1980, and more than two and a half times what they would have cut it to.

"When I first came here, I thought this was pretty sorry-looking country," Alan commented as we drove around the ranch on one of several tours we took in his old Travel-All. "Now I find myself saying, 'I remember this spot. It used to be bare ground. Now it's covered with grass.'"

Significant improvement has also been noted on the ranch's riparian areas. One old spring that had been little more than a trickle for years now extends water and riparian vegetation a half mile downstream. The present secretary of the interior, Bruce Babbitt, whose family owns a large ranch not too far north of the Orme, was impressed enough with Alan's results to ask him to give a presentation on the ranch's riparian success at one of his Rangeland Reform Hearings in 1993.

Kyle Cooper is reluctant to say whether the herd of fifty to sixty antelope that spend a considerable amount of time among the Orme's cattle represent a population increase or just indicate that local animals are spending more time on the ranch. He does say, however, that more deer are being seen on the ranch. The local Audubon Club now does yearly bird counts to help establish baseline data for the team's goal of increasing wildlife diversity on the ranch, and everyone is waiting for the first sighting of elk on the ranch as the rapidly expanding population in the northern mountains around Flagstaff pioneers its way south toward the desert.

There are human successes in this scenario, too. The ranch's owners, members of the Orme family, have started attending meetings of the strategic team, although, in the past they limited their involvement strictly to financial meetings. The advantage to bringing the two teams together, explains Alan, is that it has made coordination of overall ranch management more responsive to each group's concerns, and it adds diversity to the vision directing the management process.

Science students from a private secondary school located on what was once ranch property will be increasing that diversity even further if a memorandum of understanding drawn up by the Prescott National Forest, the Orme Ranch, and the Orme School is implemented. In 1929, the Orme family put together a small home school that grew over the years until, in 1962, it became an independent entity. Now the Orme School accepts students from all over the world. As a result of its growth, the school had become so disassociated from its rural roots that teachers were relying strictly on textbooks to teach about the rural lifestyle that was happening right outside the classroom door. The pictures those texts painted, of exploitation and overgrazing, had little to do with the thoughtful, caring management being practiced by the Kesslers.

That one-sided perspective will all change when an outdoor living classroom is put together on the ranch for the school's science students. Students who enroll in the program will help out with the ranch's biological monitoring program and, in the process, become members of the strategic team.

"When this first got started, I was just looking for a way to get people off my back," Alan confided to me one day as we sat under a cottonwood on a mat of grass beside a clear-flowing stream. The goals of the group then, he felt, were just mom and apple pie stuff; but as people became more invested in the process, they became more specific and more ambitious. "People are beginning to realize that this is about their quality of life too, not just ours," he told me. "And I've come to realize that if we want to keep making the gains we've been making, we need them to help us."

Jack Turner offered a team member's perspective. "I think we're finally getting it," he reflected. "When we saw those low figures for the success the Kesslers are getting in achieving the quality of life they want, it became clear to a lot of us that if they don't achieve their goals, we don't achieve ours."

Willow seedlings grow along Ash Creek on the Orme Ranch, Prescott, Arizona. Portions of this riparian area have become so thickly vegetated under Kessler's management that it is difficult for a human or a bovine to walk through them.

"Don't be afraid to take a big step....You can't cross a chasm in two small jumps."

DAVID LLOYD GEORGE
British Prime Minister (1916–1922)

DIAMOND C RANCH, ARIZONA

Left to right: Rukin Jelks II, Rukin Jelks III, Joe Quiroga.

CHAPTER FOUR

THE
DIAMOND C RANCH

Rukin Jelks III

ARIZONA

On Disturbance and Rest, Predators,
Making Soil, and Stretching Limits

IN ARIZONA'S SOUTHEASTERN HIGHLANDS, oak-dotted hills carpeted with blue grama grass and waving stands of cane beardgrass undulate between ranges of rugged fault-block mountains—the Mustangs, the Huachucas, and the Whetstones. Farther south, closer to the border, these grasslands become so broad and expansive that the movie *Oklahoma* was filmed among them.

Urban environmentalists and natural-history buffs have given this basin and range landscape a nickname—the Land of the Sky Islands. They use the name to give a poetic ring to campaigns for enlarging the islands of preservation within the area. Some of their efforts would restore the region's biodiversity, revitalize its riparian areas, and rejuvenate its overgrazed grasslands by removing livestock from ever-greater portions of it. Rancher Rukin Jelks III, who manages the 8,000-acre Diamond C Ranch that he owns in partnership with his father and brother, works toward those same goals, but his way of going about achieving them is about as different as can be.

The conventional wisdom tells us that the best way to restore overgrazed land is to give it a rest—the longer and more complete the rest, the better. Take the livestock off and leave them off is what scientists, bureaucrats, activists, and our own intuition tell us, and nature's processes will slowly heal the harried land and its damaged ecosystems. This concept is so firmly entrenched that most discussions of range issues treat the concepts as interchangeable. To rest land is to restore it.

Rukin Jelks rejects that convention. Furthermore, he believes that rest, or at least too much of it, can be as detrimental to a grassland as overgrazing. Instead of being an

example of the devastation one would expect to see at the hands of such a madman, the 5,000 private and 3,000 public acres of the Diamond C not only look good but compare favorably to its neighbor, National Audubon Society's Appleton Whittel Research Sanctuary, which has not been grazed in more than twenty-five years.

To Jelks, rest is just another tool. He sees nothing more magical in it than he does in fences, fertilizers, or tractors. Each tool, he knows, does some things and doesn't do others. Each is appropriate under some circumstances and inappropriate under others. All require thought and planning before they're applied.

Disturbance in the form of livestock grazing is also a tool in this way of thinking. In the same way that the managers of nature preserves, and even some ranchers, strive to keep their land as free of disturbance as possible, Jelks uses cattle to promote it on his. What some people consider an experiment too risky to take, Rukin Jelks does as a matter of routine. He is a stretcher of limits. To understand what that means you have to witness it.

On a cold day in February, Rukin, a lifelong friend of his, Bryan Childers, and I were bouncing along a dirt two-track on the way to move the carcass of a cow that had died while giving birth the previous day. The cow's calf had survived, and Joe Quiroga, the ranch's cowboy, had found it and brought it back to the barn. We had fed it, along with another orphan, before leaving the ranch headquarters that morning.

The dead cow was too close to a stock drinker for the health of the other animals, so Rukin wanted to move her to a ridge top where coyotes would catch her scent and quickly reduce her remains to piles of scat and scattered bones.

As we attended to this chore, it became obvious that Jelks has an opinion of predators that is different from most ranchers. "I don't mind the coyotes," he declares. "They clean things up. All I lost to 'em last year was a couple of calf tails." Having raised the topic, he went on to tell us about how a mountain lion had been hanging out around his cow herd a year or so ago. "I'd never seen them herd up so tight," he remembers. Because Jelks wants his cows to "herd up tight" to increase the impact they have on the land, he considers that a plus. "Hell, if they reintroduce wolves, I betcha that'd really tighten them up. And if I lose one every now and then—well,

that'd cost me four hundred bucks. You've got to pay a lot for herdin' these days. I think I'd make more money than I'd lose."

Jelks bears a slight resemblance to the late British actor Peter Sellers, even down to his mustache. The similarity is more obvious in photographs than it is in person, but once you've seen it, you see it more often. He's shorter than Sellers was, with a stockier build. In his mid-forties, Jelks talks slowly and deliberately. He chuckles again at the idea of his cows worrying about that lion as we pull up to the drinker with the dead cow next to it. Brian and I stay in the truck while Rukin jumps out and hooks a tow chain to the Blazer's trailer hitch and wraps the other end of the chain around one of the dead animal's rear legs. As we drive away with the carcass sliding over the grass behind us, he lights up a cigarette and notes that coyotes had already found the carcass and had eaten the cheek meat and tongue.

While Jelks is moving the dead animal, he decides to use the opportunity to check on the rest of his herd of 650 cattle. Spring calving is in full swing as February is nearing its end, and it would be wise to know early in the day if any other problems have cropped up. Actually, either Jelks or Quiroga check the animals daily, to monitor their impact on the range and to make sure everything is as it should be. As we clatter and lurch up to the 220-acre paddock in which the herd is grazing, I am not ready for what I am about to see, even though I have visited Jelks's ranch a number of times and am aware of the way he manages his animals.

Topping a low rise, we come into view of a pasture that encompasses a small valley and the two slopes that define it. There are cattle everywhere. They stand in clusters along the two-wire electric fence that marks the pasture's northern boundary and spread across the hillside that rises up before us. Dozens line the crest of the ridge, standing silhouetted against a February sky threatening snow. As we stop at the gate, the majority of the animals catch sight of the truck and begin bawling loudly and moving in our direction. Those that haven't seen us take their cue from the uproar, and soon they, too, are ambling toward the truck. Mother cows and newborn calves emerge from all directions, from hiding places in swales and clumps of brush. The line of brown-and-white and mottled gray bodies grows steadily longer as it trains down the hillside. With the

"Disturbance in the form of livestock grazing is also a tool in this way of thinking."

The Appleton Whittel Research Sanctuary, rested from livestock grazing for twenty-five years, is on the left. The Jelks Ranch, which is heavily grazed according to a closely monitored management plan, is on the right.

landscape virtually alive with cattle (nearly the entire herd is in view now) it becomes graphically apparent just how densely they are concentrated in this relatively small area.

As the mothers and mothers-to-be begin arriving at the fence and bunch up against it, the enormous impact the animals have had on the area also becomes apparent. Everything that a cow will eat has been eaten. Most of the rest has been stepped on. Clumps of beargrass, a relative of yucca that most cattle pass up because its fibrous leaves require a lot of chewing, have been grazed in some places to within inches of the ground. Mormon tea, a shrub that is usually just nibbled at or passed over for less-pungent forage, has been browsed as well. Cactus, yucca, and agave, which are normally grazed only lightly because of their spines and daggerlike leaves, have been consumed more heavily than I have seen elsewhere. The earth is dotted with dung, made more apparent by the shortness of the grasses that provide the land's main cover. The effect is similar to what I have seen in photographs of the plains of Africa, on land that has been heavily grazed by the huge herds of wildlife there.

Brian opens the gate, and we drive through it into the pasture where the cattle are stamping restlessly. They crowd around the truck, moving aside at the last minute as we slowly push ahead. The windshield is filled with their faces as they mill noisily in front of us. Since ranchers regularly haul feed supplements to their cattle, I assume they are anticipating a handout. Apparently, that isn't the case.

"I don't like what I'm seeing," Jelks says as he steers through the herd. "They're not eating. They're telling me they want out of here."

We drag the carcass to a low ridge and unhitch it. The animals have left us and have gone back to the fence to stand and complain. As we drive back toward the gate they crowd around us once again. A large Brahma-Hereford cross with sharp upswept horns and lopped ears always seems to be in front of us as we creep forward. Others reshuffle haphazardly just ahead of the bumper as we near the gate.

Jelks is obviously concerned. "When I moved them in here, my monitoring showed this pasture had enough forage for a fifteen-day graze," he tells me. "They've only been here for thirteen, but look at the way they've eaten things down. They've even started on the desert broom; that's

"He has just moved his entire herd in less time than it takes to smoke a cigarette, and all he did was blow a whistle."

toxic. They're telling me they want out of here, and I'm gonna listen."

At Rukin's direction, Brian unhooks the two electric fence wires that form the gate, and we drive through. Rukin parks off to the side of the opening, gets out of the truck, and pulls out the English bobbie's whistle that he uses to call his cattle. As he begins to call, the animals are already streaming through the gate, their hooves clicking and bodies rustling one against the other as they rush into the new pasture. The area they are entering hasn't been grazed in several months and the coarse giant sacaton grass growing there reaches halfway up the animals' sides. All of this has grown since the last time they grazed here early the previous summer. The grass that brushes their sides now was as short then as the stubble in the pasture they just left.

As we stand and watch the cattle spread out over the field, the cacophony of complaints is replaced by the rustle of grasses and the sound of the cattle cropping the coarse forage. Surrounded by fresh feed, the animals graze hungrily as they continue moving deeper into the sacaton. In a matter of minutes, all but a few of the 650 have left the old for the new. It occurs to me how some ranchers and most environmentalists say that the type of intensive management Jelks uses requires too much work to be practicable. He has just moved his entire herd in less time than it takes to smoke a cigarette, and all he did was blow a whistle.

The pasture the cows have entered covers only thirty-six acres. Ironically, when Jelks took over management of this ranch in the early 1980s that was how much of this land it took to support one cow (there were 220 head in two separate herds then). Even at that stocking rate the land was in extremely poor condition. In about ten years, Jelks has been able to improve the quality of the range to the point where it now supports nearly three times as many cattle and still continues to improve.

Jelks was fresh from having dropped out of a pre-med program at ASU when he arrived at the Diamond C in 1980. Without a job and with a wife and young daughter, he was looking for a place to be and a direction to take. His father, who had been a successful feedlot operator in Phoenix, had bought the ranch near Sonoita in 1972 to move himself and his family away from city society and urban decay. When Rukin once again needed a refuge, his father told him there was plenty of work to be done back on the ranch.

"When I came here to work on the place," Jelks told me, "I was just a city boy, raised in Phoenix. I didn't even know how to saddle a horse. I rode a Harley. Joe Quiroga (the cowboy who managed the ranch for Rukin's father and now works for Rukin) would take me out on horseback all day to look at the herd. We'd see about a third of it and have to go out again the next day and the next. I thought he was trying to kill me."

Jelks heard about the workshops that Allan Savory was putting on around the West and thought they promised an approach that was less "seat of the pants." He took the course and came back with a goal, a plan, and a mission.

"I want this place to be as good as it can be, with more production, more game, more bio-diversity, perennial water, and no bare ground," states Jelks. "Savory told me how I could do that. It just made too much sense to be wrong."

The job Jelks had set for himself sounded more like a pipe dream than a management plan. "The last guy to own this place had to sell out because he was starving to death," he drawled. When Jelks took over management of the ranch, the ground in some places was still nearly as bare as a parking lot. It remained so, even though Jelks's father had tried the traditional remedies for coaxing a little more productivity from used-up rangeland: plowing out the mesquites that had begun to grow in place of grass and using a harrow to rip the desert pavement of rocks and crusted soil—all that was left of a once-vital grassland. Then he seeded it to grass, including some exotics that hadn't evolved on the ranch. Besides spending a lot of money, nothing much came of those efforts.

The newly inspired HRMer knew that it would take a lot of cattle to create disturbance sufficient to get results that heavy machinery couldn't. The obvious first step to reach those numbers was to combine the two herds that had grazed separate parts of the ranch since before his father had bought it.

Jelks took his plan to the U.S. Soil Conserva-tion Service, which was advising him on manag-ing the ranch. Their response was that combining the herds would trample what little feed was left and create a disaster. Instead, they suggested bringing together portions of the herds for short

periods as a trial. Or, they hinted, if he really wanted to improve his forage, he could cut his stocking rate.

"They gave me three alternatives, all real conservative," Jelks remembered. "Move forty over for a few weeks; move sixty for a shorter period of time. Hell, trampling is what I wanted them to do. I said 'none of the above' and moved 'em all over."

What Rukin had come to believe, and was trying to act upon, was that the landscape we typically describe as overgrazed (much bare ground, eroded stream banks, declining biodiversity, and woody plants replacing grasses) is less a matter of how many animals graze than how they graze it. Grass eaters confined to a single pasture over a long period of time utilize certain areas, even certain plants, over and over again, revisiting them to eat the succulent, nutrient-rich regrowth created by their pruning. The result is that, over time, if the repeated grazings come too frequently for the plant to recover, it shrinks, weakens, and eventually dies.

Rather than being better off for being left out of this fatal attraction, the plants that aren't grazed have their problems too. Over the seasons, they accumulate leaves and seed stems, especially in "brittle" or dry areas where rot doesn't serve as a stand-in for ungulates. Having served their pur-pose, and dried at the end of the season, these now-dead appendages become a self-generated haystack that shades the ground-level growth cen-ter common to most grass plants, robbing it of life-giving sunlight. If this debris isn't removed by grazing, burning, trampling, or mowing, the plant dies. Some people call this overrest, some call it vegetative stagnation, some call it a fantasy. If you'd like to test it, try it in your yard.

When an area is subjected to this double whammy of overuse and stagnation by confining grazers there for a long period of time, not allow-ing them to move with the natural rhythms of the land, plants begin dying from both effects. Remove the grazers and you only remove half the problem. The haystack effect keeps killing plants. Force the grazers to eat less selectively and keep them moving, so they don't eat the same plants over and over again, especially during the critical time of the year when those plants are growing, and you remedy both problems—at least that's the way the theory goes. Jelks was ready to take it from theory to practice.

"Jelks has been able to improve the quality of the range to the point where it now supports nearly three times as many cattle and still continues to improve."

Before he combined his two herds, he installed the miles of fences needed to concentrate the animals more effectively in smaller pastures. Jelks uses low electric fences—two smooth wires a little more than three feet high that you can step over or step on and lead a horse across. In this new environment, the cattle were crowded together in such a way that they were forced to eat or trample even the plants that had been accumulating dried material for years. As they crowded and rushed to get to the best morsels, their hooves churned up the soil and transformed its crust into a surface receptive to seeds, and of course they fertilized it as well. Slowly in terms of what Jelks expected, but faster than his skeptics predicted, the land began to respond.

Eric Schwennesen, an international grazing consultant for The World Bank and former Agricultural Natural Resources extension agent for Cochise County, has followed Rukin's progress since he first started managing the ranch. Schwennesen monitors five sites on the ranch and says he's being conservative when he estimates Jelks's management has increased the amount of vegetative coverage there by 40 percent. "That's a very dramatic improvement, unheard of from a range management point of view," Schwennesen states. "Generally, we don't expect to see any reduction in bare ground. Just keeping it from increasing is considered a success."

To achieve his success, Jelks violated another rule of the Rest is Best paradigm: that overgrazing is purely a function of livestock numbers. As he has improved the condition of his land, he has steadily increased the size of his herd, from 220 to 350 to 450, 550, 650 and, most recently, to more than 700. Jelks will tell you that the increase in numbers did not come about because of the improvement in his rangelands, but rather that the opposite is true: the rangeland improved because of the disturbance he could bring to bear with those additional numbers. Jelks is quick to point out that he didn't just turn those animals loose on the ranch, but

"...he's being conservative when he estimates Jelks's management has increased the amount of vegetative coverage there by 40 percent."

used them according to a plan directed toward a goal and based on regular monitoring. That monitoring technique, in which he steps off the amount of ground that contains one cow's worth of forage for a day and extrapolates that figure over the pasture and then over the entire ranch, tells him he now has sufficient forage for even more cattle. Soon he hopes to have more than a thousand cattle on the same land that couldn't support 220 ten years ago.

The U.S. Forest Service has kept as close an eye as money and time would allow on the two forest service allotments that make up 3,000 of the 8,000 acres Jelks grazes. That land is subject to the agency's limitations on stock numbers and grass utilization.

"We have standards and guidelines that grazers have to follow," says Laura DuPage, range conservationist for the Sierra Vista District of the Coronado National Forest, "and utilization levels that are restricted to leave forage for wildlife." DuPage believes that Jelks has been doing a good job of abiding by forest service guidelines and, in spite of the fact that she has not been monitoring his ranch on a regular basis, she says, "I haven't seen anything that offended me." She points to two

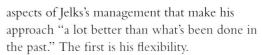

aspects of Jelks's management that make his approach "a lot better than what's been done in the past." The first is his flexibility.

"If we're having a year of poor weather, he's monitoring closely enough that he can change what he's doing," she tells me. The second aspect of Rukin's management that DuPage says contributes to his success is the way in which he continues to learn from his successes and failures. "A lot of people just take what they want from HRM and forget the rest," DuPage offers. "Rukin has made it a continuous learning process."

Jelks not only welcomes what scrutiny is paid to his operation by land management agencies but actually invites more. Because he is proud of what he has achieved and eager to show it off, he leads tours of the ranch for various groups nearly every month. Frequently, these outings are attended by agency personnel who work in other parts of the state. They're also attended regularly by environmentalists, scientists, and other ranchers.

Usually, these outings incorporate a pass through the Audubon Research Sanctuary, which has been rested from cattle grazing since 1968. (The shortest way between the Diamond C's two extremities is by way of a road that passes through the sanctuary.) In October of 1992, I helped organize such a tour, which visited parts of a nearby Nature Conservancy preserve as well as the Audubon sanctuary and Jelks's ranch. Serving as facilitators for the trip were Jelks and the directors of those two other facilities. A number of the top grassland scientists in the Southwest and people from various government agencies, writers, ranchers, members of Earth First!, and staff members of the Arizona Nature Conservancy made up the rest of the forty-six people who attended.

On a piece of land that had been rested from livestock grazing for a quarter of a century, Jelks and Tony Burgess, program director of the University of Arizona's Desert Lab near Tucson, argued over the health of the soil and the plant community it supported. Jelks pointed out that the collection of grass plants that the area supported were of low vitality and were perched on pedestals of soil separated by the rocky evidence of erosion. Burgess, who has been very critical of the harm grazing has brought to the Southwest, interpreted what he saw as plants sloughing off debris and waste to create hummocks of soil where past grazing practices had caused most of it to wash away. "It takes thousands of years to make an inch of soil," he said, as he dug in the earth and produced evidence of algae to show that the soil did indeed have life in it.

When we reached Jelks's ranch, the first thing that became obvious to everyone was that the grass, which had barely reached above our sneakers on a similar dry mesa top on one of the preserves, reached above our belts here. When I tried to photograph the turf in a stand of native cane beardgrass, aiming my camera straight down, I couldn't focus because the grass seed stems were nearly touching the lens. To take the photograph I had to step back into a shorter stand of curly mesquite. The area where all this grass was growing had been grazed less than a year before. All the grass here had grown since then. Photographs of the same place taken ten years previous to our visit showed more bare dirt than grass.

While most of us were walking around being amazed by the tall grass, Burgess was poking around in the soil with his digging knife. As he lifted a handful of dirt from under the grass and crumbled it in his fingers, he noted that it was still moist though it hadn't rained in weeks. Flipping over a piece of cow dung, he discovered abundant evidence of termites and other subterranean life, evidence that he had not found on the sanctuary. Turning to Jelks, Burgess said, "I don't know what you're doing here, but whatever it is, don't stop. You should be very proud."

Because of the controversial nature of the issues surrounding cattle grazing in the American West, I was reluctant to quote some of the things Burgess said that day for an article I was writing about the outing for an environmental magazine. I didn't want to get him in trouble. So, I called him up, read some of my notes back to him, and asked if he had meant what he said about the success of Jelks's management. While he still expressed concern about the effects cattle can have on a desert grassland and dismay at how confrontational the outing had become at a few points, he did say, "I'm convinced that this sort of livestock management can work if applied by someone as dedicated and diligent as Rukin Jelks, and you can quote me on that."

"I don't know what you're doing here, but whatever it is, don't stop."

Scientists, environmentalists, ranchers, federal land managers, and wildlife agency staffers inspect the fruits of Rukin Jelks's intensive management in an area where short bursts of heavy impact are alternated with longer periods of rest.

*"One can own, either rightfully or fruitfully, only
those things—and only so much of a thing—as one can
come into some intimate relationship with. One cannot really
own any land to which one does not in turn belong, and
what is true of land is true of everything else."*

JOSEPH WOOD KRUTCH
from The Desert Year

RANCHO DE LA INMACULADA, SONORA, MEXICO

Left to right: Don Jesus Jimenez, caporal *(foreman), Francisco Valdez, Fermin Ruiz,
Ivan A. Aguirre (owner), Jesus Antonio Solis, David Valdez*

CHAPTER FIVE

RANCHO DE LA INMACULADA

Ivan and Martha Aguirre

SONORA, MEXICO

On Goals, Timing, Bulldozing the Desert,

Impact, and Restoration

WHERE THE HELL IS IT?"

"I don't know. The last time I flew down here, the ranch stood out like a sore thumb. The whole place has been bulldozed. You could see it for miles."

Three of us—Jay Dusard, Rukin Jelks, and I—were circling in a light plane above the Mexican state of Sonora, about halfway between the Sea of Cortez and the highway that parallels its coastline about sixty miles inland. We had come to visit Ivan and Martha Aguirre at their *rancho* on the Rio de la Inmaculada, seventy-five miles north of Hermosillo and about the same distance south of Nogales on the United States–Mexico border. Below us, an eclectic landscape of glass-sharp volcanoes, rounded hills, and sandy smooth riverbeds slipped by at 140 knots per hour. The land was speckled with mesquite and cactus scattered randomly across a background of mostly bare dirt. A few small towns were clustered along the roadsides. In the hinterlands, which means virtually everywhere, small ranchos persisted where they had been placed by opportunity and maintained by diligence. All but the steepest and most rugged reaches of the countryside were crisscrossed with an indecipherable hieroglyph of roads in every imaginable condition.

Several of the ranchos we flew over had been bulldozed to clear them of the wiry brush and bristling cholla that inevitably invade grasslands that are grazed relentlessly. Many were so bare it was easy to assume that the clearing had happened recently, but the dryness of the region and associated slow rate of recovery were a more likely explanation. Jelks circled lower, hoping we might be able to recognize the Aguirres' red brick hacienda, with its central courtyard and airstrip among the bulldozed ranches. No luck.

We had lots of gas, so that wasn't a problem. Still, our frustration mounted, at least mine did. How do you get unlost at 8,000 feet? We couldn't stop and ask directions.

"There are the stacks of the power plants along the Sea of Cortez," Jelks shouted above the engine noise. "According to the bearing we're flying, and our distance from Hermosillo, the ranch has to be right ahead of us, over that low string of hills."

Jelks held the map on his knees as he measured distance on it and plotted our course. He reached to touch the controls only when the plane tilted precariously to one side or the other. When we passed over the hills and looked down, there was the ranch. Because we were retracing the course we had followed from Nogales, we knew we had missed it at least once. Now it became obvious why. Instead of bare dirt cross-hatched with bulldozer tracks, La Inmaculada was covered with vegetation—green grass, shrubs, and trees.

As we circled to land, willowy limbs of baccaris trees swayed below us in the breeze, their dried blooms and seeds flashing silver in the sunlight. Mesquites stood out as polka dots of dark green among the grass that had taken on a light olive drab color as it cured in the dry, hot days of the Sonoran autumn.

With its broad grasslands peppered with wispy-leaved trees, the landscape reminded me of the flight scene in the movie *Out of Africa*. In that movie, Robert Redford as Denys Finch-Hatton takes Meryl Streep as Isak Dinesen on a flight above the plains of the Serengeti. John Barry's unforgettable soundtrack is playing in my head.

Well into our approach, three horses step out of the trees onto the landing strip. Jelks snatches the plane out of its steep descent, and, as he increases the pitch of the propeller, the added stress causes the engine to roar in complaint. We circle and come in for our second approach. Someone is chasing the horses clear of the landing strip. Jelks sets the plane down smoothly. A crowd of kids, Ivan, Martha, and members of the housekeeper's family, who also live in the hacienda, come running and shouting as we taxi up to the tie-down.

The name of the ranch, La Inmaculada— the Immaculate One—refers to the Virgin Mary. It dates back to colonial times, when the Spanish gave Biblical names rather than descriptive ones to places on the alien landscape of New Spain. In this case, the name refers to the rio and the plateau it drains, but in an ironic way it fits the ranch as well (at least it did when Ivan Aguirre's father, Mario, bought it in 1975). Then, Rancho de la Inmaculada was to be the Brave New Ranch, the most progressive in all of northern Mexico. Señor Aguirre enlisted all of the best minds, from Mexico City to Texas Tech, to make sure that goal was achieved.

"We reinvented our valley according to the most persuasive ideal given us by our culture," Ivan relates, fourteen years after the completion of the ranch's rebirth, and nine years after its death. "And we ended up with a landscape organized like a machine for growing crops and fattening cattle, a machine that creaked a little louder each year, a dreamland gone wrong."

In 1980, the Brave New Ranch celebrated the completion of the "reinvention" with a gala in Hermosillo, the nearest city and hometown of the Aguirre family. That grand occasion was attended by the governor of Sonora, the local archbishop, and dignitaries from both south and north of the border.

The transformation of 10,000 hectares of the valley of the ephemeral Rio de la Inmaculada into a cattle-raising machine had started with the stripping away of all the native vegetation—the saguaros, the mesquite, the native grasses, and the shrubs. "We wore out three bulldozers," remarks Ivan, "three big ones."

The bulldozers uprooted those venerable residents and scraped and piled their broken remains into windrows four feet high, forming a pattern of chevrons across the land. These low dikes, made mostly of soil scraped up with the trees, were designed by the best science of the day to collect what rain fell on the ranch and keep it from escaping. In the process of creating this grid-work, much of the shallow, braided bed of the rio, the work of millennia in this arid land, was obliterated. The river was thus left to wander aimlessly across the ranch when it did flood, bits and pieces of it becoming trapped by the dikes and held back to nourish the cows.

After being bulldozed, the land was separated into pastures, cropfields, and feedlots. Ten wells

"Ivan refused to sign on to the sale of the ranch, choosing to incur the wrath of his family rather than to give up on his dream…"

This photograph was taken just after the site had been grazed in July 1989. Notice the bare dirt around the large bush.

The same place in September 1992. In spite of the fact that the land is grazed periodically by 3,000 cows, what was bare dirt is covered by a dense stand of perennial plants.

with nature than playing with his friends. To Ivan, the ranch promised a way to live his whole life with animals and nature. He attended college at Texas Tech to study economics and range management to prepare himself for this life of dreams. But while Ivan was away in a foreign country learning how to operate the Brave New Ranch, his family was coming to realize that the immense cow-raising machine they had created was really nothing but a monstrous grave digger. And it was slowly but surely burying itself, and them, in debt.

Ivan's family has a long history in business, a successful history. It owns a number of other businesses besides the ranch, including a car dealership and a Japanese restaurant "with a sushi bar" in Hermosillo. As diesel prices rose from one peso per liter to 500 pesos, and inflation cut deeper into profits, the best of business senses could not keep the Brave New Ranch afloat. Creaking and shuddering like a great wounded ship, it began to sink. As it foundered, it threatened to suck the family's entire enterprise into the depths with it, as a sinking ship pulls down the lifeboats of those trying to escape. And in the middle of all this, Ivan's father died.

"It was the most terrible time in my life," Ivan remembers. "The family was coming apart. We were facing bankruptcy. My own brothers were trying to put me in jail!" Ivan refused to sign on to the sale of the ranch, choosing to incur the wrath of his family rather than to give up on his dream of living on the land. During this time of trouble, several remedies were tried. The irrigation equipment was sold. The ranch was leased to another family to be used for feeding stockers—steers turned out upon the land to grow from calves until they were ready for the feedlot. The Aguirres tried feeding stockers themselves.

were drilled to tap a deep aquifer of cool water, which has persisted under this desert since the last ice age 10,000 years ago. Massive irrigation systems were installed, first with water cannons and later with center pivots, such as the ones used to water the grain fields of the American Midwest. Underground silos in the form of deep trenches were gouged into the earth to store the feed harvested from the irrigated fields. Exotic species of grass were introduced into the pastures to complete the transformation from native plants begun with the bulldozers. So it was that 24,000 acres of Sonora were thus transformed from desert into an agricultural industrial park of motors, pipelines, feedlots, and exotic plants.

"All this to raise food for cows," marvels Ivan. He was a young man in high school when all this began in the mid-1970s. He grew up in Hermosillo, a city kid with a two-acre backyard where he raised quail and spent more time being

In the midst of his despair, Ivan attended an agricultural conference in Hermosillo. Most of the programs were routine, about animal genetics and exotic forage plants—quick fixes that the Aguirres had already tried. But one talk was different. It was given by an American who spoke through an interpreter. "Most of the 400 people there got lost," Ivan said. "They missed the speaker's point. But I didn't. I caught the interpreter's big mistake."

Everyone thought the speaker was talking about putting more cattle on the land and then moving them around a lot. It sounded like smoke and mirrors—more cattle and the land would get better. But it sounded like it would make more money, and that more than anything is what they all wanted to hear, even Ivan. Ivan heard something different, though. He heard the speaker say that time was the key. Even the interpreter missed that. Time was the key to grazing animals in harmony with the natural cycles of the plants and the land, to treating a ranch as a whole ecosystem and planning management practices to fit into that whole instead of forcing the ecosystem to fit management. Ivan found that the message reached into his despair and touched him with hope. "I was crying like a little kid when I went back to Martha and said, 'You must hear this man. Finally, something is making sense.'"

The first thing that strikes one about Ivan is his enthusiasm; the second is that it is rooted in innocence. Martha, who majored in economics at Texas Tech, is a perfect compliment. She is self-assured but warm, confident, and competent. She is gracious and down to earth. She and Ivan talk about this time in their lives with great emotion. To them it was a rebirth.

The man the Aguirres had run into was a representative of the Holistic Resource Management (HRM) Center in Albuquerque, New Mexico. Ivan had done a study on HRM as part of his graduate work at Texas Tech and concluded that it didn't work. This time, after he and Martha reinvestigated HRM and began applying its principles and processes, the couple became totally immersed in the holism and goal-directed planning it advocates. A copy of the HRM planning model, translated into Spanish, remains tacked to the wall of the Aguirre hacienda, under the porch roof, outside the office. It is visible from everywhere inside the courtyard. Copies of the model, or sections of it, are taped to the walls of the cow

camps, above the rough wooden tables where the *vaqueros* and their families drink coffee from the pot kept perpetually steaming on wood cookstoves made from fifty-gallon oil drums. They all participate in planning meetings now. Ivan's brothers and sisters have even returned. Ivan has gone from taking classes on HRM to giving them. His neighbors, who at one point were willing to believe that he and Martha had been brainwashed by a cult, have changed from avoiding him to coming by and asking questions.

Ivan is eager to show us his land as he shows it to them, to get us down on all fours, so we can look at it closely. Sounding like Don Juan, Carlos Castaneda's mythical desert *brujo*, Ivan tells us this is the only way to "open our paths of perception." Before we look at the land, however, we must hear about his plan and about his goal. In HRM, the way Ivan practices it, the goal is primary. "Without a goal," he declares, "you have nothing."

We gather on the porch, in front of the icon-like display of the HRM planning chart. "Before we could define our goal," explains Ivan, "we had to define the whole." The "whole," which is the subject of the Aguirres' plan, encompasses the 24,000 acres of the ranch, the Aguirres, and their cattle, that's obvious. But it goes beyond the obvious. It includes the vaqueros and others that work regularly on the ranch, their families, and the neighbors and migrant workers who rely on it for part of their income. The number of people directly affected by the plan totals fifty-five.

The whole also encompasses the plants and animals that live on the ranch all or some of the time, from the mule deer that are returning with the native vegetation to the microbes that promote the decay of organic material, thereby serving as a foundation upon which the rest of the ranch's biomass depends.

The goals for La Inmaculada were put together by the ranch's management team, which includes the vaqueros and other workers who make their life on the ranch, as well as Ivan's family. Martha has been trying to get women of the community involved as well, but that cuts hard against tradition. "They will come," she states simply.

The goals put together by this team are at once sweeping and specific. Ivan reads the list, reverting to Spanish when the English word refuses to come, "To advance ecological succession and increase biodiversity," he reads. "To enjoy free-flowing water and the cottonwoods,

"Time was the key...to treating a ranch as a whole ecosystem and planning management practices to fit into that whole instead of forcing the ecosystem to fit management."

willows, and other types of riparian trees that past generations have told us once lived here. To cover the maximum extent of soil with grasses, herbs, shrubs, and trees. To enjoy physical, emotional, and spiritual health. To increase profits economically and socially from administering flora, fauna, and cattle. To plant, harvest, and gather our own food."

Some of the goals contain elements that seem to be contradictory: flora, fauna, and cattle. Cattle are present in virtually all of them. To some, they taint the whole process, turning what sounds like an ambitious plan into an environmental red herring. To Ivan, they are agents of restoration.

Having been introduced to the goals, we climb into the bed of a pickup for a tour of the ranch. The truck bounces over the rutted clay roads heading for the place where Ivan's herd of 3,000 mother cows and bulls, along with their offspring, are grazing. As we clatter over the ruts, dodging the limbs of mesquite and baccaris that spring off the cab and slap into the truck bed, white-sided jackrabbits materialize like ghosts in the brush ahead of us and go bounding off. Evolution has equipped these Sonoran Desert hares with white-shafted, gray-tipped hairs on their haunches and flanks. They are able to erect these at will, drawing pursuers' attention to the rear of their body and creating enough of a distraction to cause a miss in a game where accuracy calls for leading the target. I watch as several jacks complete their transformation from inconspicuous gray bump to white-sided, big-eared prominence and lope away.

Jackrabbits are just one of many natives that are returning to the ranch now that management efforts to remove the native vegetation have stopped. "My father used to burn so much mesquite that the neighbors called him Nero," Ivan jokes. Under Ivan's management, mesquite, baccaris, palo verde, and other native trees and shrubs now are common throughout the ranch. Even the slow-growing saguaro can already be found in heights of up to ten feet. While some ranchers might consider the return of these woody species a liability because they compete with grass for available moisture, Ivan considers it an asset. His cattle have replaced wildfire as a source of disturbance sufficient to maintain the grasslands and keep the trees from taking over the ranch, and the sweet beans of the mesquite tree

provide a flour that the families living on the ranch use to make tortillas. Mesquite bean flour is so sweet the Aguirres use it as a substitute for sugar in pastries and sometimes in coffee. Ivan says he is working to market it to health-food stores in the United States.

Another action La Inmaculada has taken to turn disaster into asset has been to exhume the carcasses of the mesquite trees that were bulldozed into windrows and turn them into charcoal to be sold north of the border. As we drive across the ranch we see the wispy smoke of one of La Inmaculada's *carboneros* camps. Here, mesquite hulks are baked into charcoal in earthen pits. The ranch ships twenty tons per week of Chapparral-brand charcoal into southern California.

We stop briefly at the carboneros camps, but our main destination is one of the ranch's five cow camps. These outposts are located at central water points where a number of pastures converge. Constructing the pastures in this way makes for efficient placement of water sources and enables the vaqueros to stay in one place as the cattle are moved through several steps of the grazing rotation. Each camp's sole permanent structure, except for a corral and watering trough, is a small cinder-block building that serves as a bunkhouse and kitchen.

The back room of the camp we are visiting has been blown off by a tornado. A plastic bucket covered with cheesecloth sits on the bare slab holding a batch of farmer's cheese. This simple and easy-to-make cheese is another of the food items produced on the ranch for the people that live there. The vaqueros and their wives sell the excess to neighboring families as a source of additional income.

As we pull up, two women are inside the cinder-block building keeping the coffee hot. Children play in the yard. Saddles hang under the porch roof. Don Jesus, the venerable old cow boss, and his vaqueros are there to meet us. Don Jesus has been at this business a long time and the inevitable spills from the back of a cow pony have taken their toll. He supports himself by leaning on whatever is handy whenever he is off a horse.

Out at the water point, cattle come and go in a steady, bawling stream, drinking the water drawn from the only one of the ten wells drilled to irrigate the Brave New Ranch that is still being used. During wet times of the year, water is drawn from a surface water well on the banks of

"His neighbors… changed from avoiding him to coming by and asking questions."

This early spring photograph shows that native plants such as saguaro and mesquite are returning to Rancho de la Inmaculada as well.

the rio, thus avoiding further depletion of the underground aquifer.

The cattle of La Inmaculada are kept in one large herd to concentrate their impact, and few ranchers concentrate impact as much as Ivan Aguirre. As we walk out into the paddock, we scuff our feet through soil that resembles what you find in a tilled garden. As a matter of fact, Ivan actually uses his herd of 3,000 cattle as tillers, grazing them for half a day in a field of fifty acres to prepare it for the planting of tepary beans and Indian corn, native plants once cultivated by indigenous peoples. The soil where we stand now, near the water point, is pulverized. Cattle dung and grass litter are churned into the mix, along with limbs and twigs busted from shrubs and trees by the continuous passing of the cattle on their way to water, creating a loose mulch in some places up to six inches deep. To our left, two immense Brahma bulls paw and churn the soil as they lurch into one another grunting and straining. We name one with crumpled horns Sumo because he is so huge and because his fighting style reminds us of Japanese wrestlers. I look at the devastation caused by all this activity and wonder: If this is what a grassland looks like after 3,000 cows,

what would it look like after 100,000 buffalo or 1,500,000 wildebeest? As far as we know, this area did not have great herds of anything, but it responds as if it remembers.

Rukin Jelks, who knows something about disturbance, reaches down and picks up a handful of pulverized dirt. "Do you know what would happen to this if it rained?" he asks. Massive erosion, I think, but before I can reply he continues. "It'll soak up water, lots of water, and soil around here isn't used to being wet. When it gets wet," he says, "it explodes with growth."

In the midst of all this disturbance, if you get down low to the ground and look across it you can see the clumps of stubble that tell us there are grass plants here. Time, Ivan reminds us, is the key to their recovery. He will move his cattle before they cause permanent damage to those plants. Across the fence in an adjoining pasture, just as close to water and therefore just as devastated a year ago, the grass stands to our armpits.

Near another water point, a cow path is being filled in by recently sprouted annuals. At yet another, where the ground is so bare that the distance from a monitoring point to the nearest perennial plant is twenty-nine centimeters, more annual seedlings lift saucer-sized pieces of soil crust as they reach for the sun. Ivan assures us perennial plants are slowly reclaiming areas like this, and he has before-and-after photographs to prove it.

Depending upon your point of view, it is possible to see what is happening here in a negative as well as a positive light. When I described what I saw on La Inmaculada to a biologist friend back in Flagstaff, she asked "Are the grasses exotic or native?" When I answered, "exotic," she frowned. The most common grass growing on La Inmaculada today, is buffelgrass, the same import from Africa that Ivan's father planted in the 1970s. Today, however, it exists in a mosaic with native grasses and shrubs. In some places, natives have been the more aggressive colonizers; in other places, they have not. Where the ground is bare, Ivan still throws out buffelgrass seeds along with salt to attract the cattle and concentrate their impact, so they will plant the seeds. When I asked Ivan if the native species that evolved here would ever replace buffelgrass if allowed to compete freely, he replied, "Buffelgrass is now a native in northern Sonora." That is so, he says, because what was done to La Inmaculada during the 1970s has been done to hundreds of other ranches here. Millions of acres have been seeded to this aggressive exotic, which now is spreading on its own. Ironically, a type of native buffelgrass also lives here. We find a few plants hiding among some baccaris trees in a shady, moist area. "Why hasn't the native buffelgrass prospered?" I ask Ivan. He doesn't know.

While the native buffelgrass still defers to its exotic cousin, baccaris is a native that has successfully returned to La Inmaculada. These are the trees that waved so beautifully in the breeze as we flew in for a landing. Baccaris, or *romerillo* as it is called in Mexico, is a member of the seepwillow family. On neighboring ranches, this large bush occurs only in the vicinity of the river or near stockponds or puddles created by dikes. On La Inmaculada it occupies extensive areas of the ranch. To Ivan, this stands as testament to the fact that the vegetative mat his management is creating is serving as a more effective retainer of moisture than the bare dirt it is replacing. (Maybe that will eventually bring back the native buffelgrass, too.) The fact that baccaris can be poisonous to cattle at certain times of the year would have other ranchers planning a way to get rid of it. "No way," says Ivan. "It is part of my biological capital."

Any visit to La Inmaculada seems to include many pleasant moments spent around the Aguirres' large, circular dinner table. Martha is an artist at cooking on the wood cookstove, preparing delicious meals at a moment's notice that taste even better when spiced with the seasonings that sit on the lazy susan. (Watch out, those are really hot!) One evening, after a marvelous meal of *machaca* (dried, pulverized beef), mesquite bean flour tortillas, tepary bean refritos, and farmer's cheese, we lingered around the table telling outrageous jokes. Most were notable for their badness, but all achieved the desired effect of promoting an atmosphere of good-natured, cross-cultural camaraderie. Afterwards, we retired to the porch, which extends entirely around the inside of the Aguirres' compound, looking inward on the courtyard. This peaceful, embracing environment, ornamented with bougainvilleas and cactus, is mowed by desert tortoises that have dug burrows under the walkways.

Seated on a wooden bench under the stars that reclaimed the sky after the generator was shut down, I asked Ivan, "What would you say if I told you that I don't believe you can graze cattle on arid land without harming it; that you damage the land by grazing it, you don't help it?"

Without hesitating, he replied, "I would ask, 'What are your goals?'"

A bit caught off guard by that response, I answered, "Clean water, a healthy ecosystem, wildlife…"

"Then, I would tell you that it works," he responded. And even though it was dark, I could tell he was smiling.

"As far as we know, this area did not have great herds of anything, but it responds as if it remembers."

*"Eat and respect. Cultivate your garden, on your
hands and knees. Eat weird things like lions and hawk;
taste the wild; save road kills. Eat a cow from a good
rancher; know who the good ranchers are."*

STEPHEN BODIO
from Struck With Consequence

MILTON RANCH, MONTANA

Left to right: Dana and Bill Milton, Rona Johnson, Howard Johnson.

Not pictured: Max Milton, Wilbur Wood.

CHAPTER SIX

THE
MILTON RANCH

Bill and Dana Milton

MONTANA

On Trust, Zen, Finances, Organic Meat,
and a Grass Culture

WHEN I VISITED Bill and Dana Milton's sheep and cattle ranch north of Roundup, Montana, in July 1993, the ranching industry's political spokesmen were predicting the death of the western livestock industry. Secretary of the Interior Bruce Babbitt had canceled the wool subsidy and was working hard to enact a grazing reform package that would more than double the fees ranchers pay for grazing their livestock on public lands. In addition, those reforms would shorten the tenure of grazing allotments and transfer ownership of water rights on public land allotments to the government. If you were a rancher, things were looking bleak.

The Miltons' response to all that was, "We're talking great opportunities here."

If you conclude from that comment that the Miltons aren't typical ranchers, you would be more right than you know. Of course, there is no such thing as a typical rancher, any more than there is a typical writer, or a typical African-American, or a typical environmentalist. I mention this because it's a truth that we forget all too easily. The types of gross generalizations we readily condemn when they're used to refer to ethnic, sexual, or cultural groups are still pretty much business as usual among much of our society when the butt of the reference is people who share an occupation or lifestyle. Nevertheless, it's no more legitimate to call someone a welfare rancher than it is to call them a dumb blonde.

Bill and Dana Milton are outstanding examples of the diversity that exists among the people who raise livestock in the American West. At the same time, they're excellent examples of the diversity to be found among people who call themselves environmentalists. The Miltons have not only stood on both sides of that fence but they've also spent much of their adult lives trying to build bridges across it.

On the wall in the den of their renovated ranch house, Bill and Dana's wedding portrait shows them in full West Coast counterculture garb, with lots of long hair and colorful clothing. Dana is dressed in a sprightly gown with spring flowers woven into her hair and spilling from her bouquet. She looks like a model for one of the Alphonse Mucha posters that people had thumbtacked to their walls back in the 1960s and early 1970s. The person performing the ceremony could have stepped right off the cover of a Grateful Dead album. The celebration looks like a druidic rite, because it was; Bill looks like anything but a cowboy, but he is.

Bill grew up on a cattle ranch in northern Montana. After his parents split up, he divided his time between Montana with his father and San Francisco with his mother. After a year at Montana State, he transferred to the University of California at Berkeley, going back to Montana during summers and vacations to spend his time in the cow camps.

It was in Berkeley that he and Dana met. "I was attending the same boarding school as Bill's brother," she smiles, "and I just fell in love with this tall, handsome Montana cowboy in his turquoise-and-orange cowboy boots."

When Bill's father died in 1972, it was stipulated in his will that the family ranch be sold. A buyer was found, but when Bill discovered that their aim was to subdivide, he walked into the San Francisco headquarters of the Nature Conservancy and asked them if they wanted to buy a ranch. After they found out the particulars, they decided that they did. The Nature Conservancy had to take the case to court, and in the end the ranch brought the family less money than it would have if it had been sold to developers. But Bill got his brothers to agree, and the deal was made.

With Bill's ties to the family ranch cut, he and Dana became deeply involved in environmental issues. In the Bay Area, Bill helped prevent a freeway ramp from being built through habitat critical to an endangered bird, the clapper rail. After marrying and moving to Montana, Dana did a stint on the board of the Northern Rockies

Action Group. Back on his home turf, Bill served on the board of directors of the Northern Plains Resource Council, a group that has long stood as an example of environmentalists and ranchers working together. During that time, he took the opportunity to try his hand at cowboying again and found his love for the ranching way of life rekindled. By 1978, the Miltons had bought the Milton Ranch north of Roundup in the old Mussleshell River bison country. It consists of 14,000 acres of grassland—9,000 private acres and 5,000 BLM grazing allotment acres.

My visit to the Milton Ranch came just as a three-year drought had ended. A year's worth of moisture had fallen in the preceding few weeks, and the explosion of growth that had been detonated by the gentle grace of rain lent an almost unnatural aura to what was otherwise a study in Andrew Wyeth starkness. Weathered sheep barns and rusting farm implements floated in a waving sea of blue grama and green needle grass. Mosquitoes were so thick I had to walk with my mouth closed to keep from swallowing them.

I arrived in the morning, and though I was expected, everything was business as usual. The monthly financial team meeting was about to begin, and Bill was on the phone with a photographer from the *Billings Gazette*. The photographer was making an appointment to finish a feature on the Miltons' innovative management practices, and Bill was telling him tomorrow would be a good day since he would be showing me around the ranch, too.

As members of the financial team started to arrive, the radio was broadcasting a news report about Interior Secretary Babbitt making a swing around the West, conducting hearings on the ill-fated first of his two grazing reform packages. Though the increase in fees those reforms would bring wasn't a primary topic of conversation, it was obviously on everyone's minds. As background, it added an element of destiny to the financial discussion, even more so since, as Bill Milton put it, "Things are more intense this year because we're in a bit of a bind."

Bill assured me that he had blocked out enough time for us to tour the ranch to discuss his management practices and his efforts at bringing urban and rural environmentalists together. First, however, he urged me to sit in on the financial meeting. "This is the guts of it all," he declared. "It makes all the rest of what we're

"Money is an implement for creating a quality of life that not only nurtures the people associated with the ranch but extends into the surrounding community as well."

The Miltons have set aside this reservoir on their ranch as wildlife habitat. Notice the waterfowl nesting platform along the shore of the island. The Milton Ranch is also managed as a walk-in-only hunt unit to avoid vehicle damage to the environment.

doing work. You really ought to be part of it."
For a day, at least, I was a member of the Milton Ranch financial management team.

That team consists of Bill and Dana, Bill's brother Max, who is part owner of the ranch, and ranch manager and hired hand Rona Johnson. We were to meet in the kitchen of the house the Miltons have redecorated and expanded into a home beautiful enough to have been featured in *Country Home* magazine. We pulled up chairs around the table, which was bordered on three sides by windows looking out on grassy fields, barns, lambing pens, and the road to town. Dana sat ready at the computer. The rest of the team had their notes spread in front of them. The coffeepot was full.

The discussion was wide ranging—about much more than dollars and cents, hay and tractors, cows and sheep. More than anything else, it was about life. In the planning model that the Miltons use, money is a means, not an end. It stands with trucks, sheepdogs, livestock, and labor as an implement for creating a quality of life that not only nurtures the people associated with the ranch but extends into the surrounding community as well. It also provides the means for creating a landscape goal set for the ranch by a larger, more diverse team. That larger team includes government land managers, neighbors, and a poet.

Tools and goals, lives and plans all became interwoven in the discussion. Each team member took a turn "sharing." Bill and Dana said that a lot of their time has been spent away from the ranch recently. Throughout the summer, Dana had been shuttling their three kids and teammates to weekend swim meets around the state. "It has been taking most of my energy," she said, beaming. One of her two sons was in the running for a state championship.

In addition to being a mother and part of the ranch team, Dana is an activist—a mover and shaker in both the spheres of business and community. She's strictly no-nonsense, blunt and direct. Her mind is as quick as her eyes are bright.

To help market the wool that the ranch produces, Dana has put together a hand-spun yarn company that employs members of a local religious community, the Hutterites, as well as other neighbors. This cottage industry produces yarn that is sold internationally under the label Ramboula. Named after the breed of sheep the Miltons raise, it has a sales and shipping outlet in Roundup.

Another of Dana's extracurricular activities is a continuation of the environmental activism that she started in the Bay Area and stands as an expression of her commitment to the future of rural culture and economy: She is chairperson of the board of directors and one of the founders of the Montana Land Reliance (MLR), an organization that negotiates conservation easements on private lands. The lands that the MLR seeks to protect are those "that are ecologically significant for agricultural production, fish and wildlife habitat, and open space." Emphasis is on lands that are under the greatest pressure from subdivision and development. As a result of the MLR's efforts, more than 103,300 acres of "America's most prized rivers and streams, farm, ranch and timbered lands" have been permanently set aside from subdivision and development. These lands include more than 200 miles of streamsides and riverbanks, more than 26,600 acres of prime elk habitat, and more than 3,600 acres of wetlands. Of these lands, 44,000 acres are in the Greater Yellowstone Ecosystem.

When it came to Bill's turn to share, he admitted that he, too, had been concentrating on things other than the ranch. "I haven't spent much time on the ranch in the last thirty days, so I feel a little weird and rushed," he told us. "I need to pound some steel and look at animals." Chop wood, carry water. Bill is a Zen rancher, or rather, a rancher who practices Zen. One of the things he does beyond the ranch is to attend retreats to meditate and study eastern philosophy. He recommended a book he has just read, *The Tao of Negotiation*. It has helped him learn to communicate better with Dana, he said.

Bill's brother Max lives in Helena and is also heavily involved in environmental issues with an organization named A Territorial Resource (ATR), based in Seattle, Washington. He brings a different set of environmental concerns to the discussion. Max is worried about the Wise Use movement and the revolt against environmental regulation it foments by casting those regulations as a taking of private property rights. "I've heard some real horror stories," he said, "about people buying private land and clear-cutting it to remove all the timber before regulations prohibit it." Max has an opinion on grazing fees, too. "You know they ought to be higher," he said. "Maybe not eight dollars like they're talking about."

Rona Johnson, the ranch's only full-time hired hand, brings the conversation back to the here and now. Speaking in her rich Welsh brogue, she reported first on the ongoing program to train a new group of guard dogs to protect the ranch's band of sheep. (Sheepherders call them bands.

"Bill is a Zen rancher, or rather, a rancher who practices Zen."

The Bible calls them a flock.) Guard dogs don't herd sheep, as Border collies do; as puppies, they bond to the sheep as kin and protect them from predators, just as a wolf would protect a member of its pack. Rona explained that a breakdown in the guard-dog program became obvious recently when she went to check on the sheep and found one of her Great Pyrenees dogs sharing a meal of fresh mutton with a coyote. Rona told the team that she was in the process of breaking in a new crew of Afghan AkBars that were reputed to be more aggressive. "They're more likely to eat the coyote than the sheep," she chuckled.

Rona went on to say she was getting more excited about an idea that had been discussed at the last meeting—the idea of raising organic beef and lamb on the ranch. The Milton Ranch was the first in Montana to have its meat certified "natural," which means it is raised without hormones or antibiotics. Known as O Bar M Meats, that program was another of the innovations the management team had come up with to make sure the ranch remains economically viable and socially responsible. Now they were thinking about taking the next step: raising meat that could be certified organic. To do so, their animals would have to be raised not only free of hormones and antibiotics but without consuming a bite of food that was treated with pesticides or fertilized with anything but organic fertilizer. The best way to do that, they had decided, was to become totally self-sufficient, growing all the feed their animals consumed from cradle to grave, from lambing pen to market.

Raising animals to these standards is no easy matter. In order to get a handle on it, some companies now track all their feed by computer. For the Milton Ranch, money to develop the program might be the deciding factor. The bottom had fallen out of the wool and sheep market. "And a couple of years ago sheep were so profitable we were thinking about going out of the cattle business," Bill marveled. With sheep prices down, the computer spreadsheet Dana was working on as the meeting progressed refused to come up with numbers in black ink rather than red. "We're still short," she said, as she made a few more changes in the spreadsheet.

"How much?" someone asked.

"Thirty thousand," Dana replied.

"The part of my wages that is paid in livestock could be attached to the ranch's performance," Rona suggested.

"But in a bad year that means you would end up getting less," Dana responded, "and you work harder in a bad year than you do in a good one."

"That's business," replied Rona.

Other alternatives are discussed. Currently, a nonnative plant named leafy spurge is spreading across the West. Spurge is considered a noxious weed because cattle won't eat it and because it crowds out both native plants and more acceptable exotics. Sheep, however, thrive on the stuff. "I think my lambing rates go up when my sheep are on leafy spurge," Bill Milton joked.

Renting out the sheep for spurge control is discussed, as well as saving on equipment costs by getting a neighbor to hay the meadows. A considerable amount of money could be saved by letting the sheep bear their lambs on the range instead of in the more controlled and protective environment of lambing pens. None of the alternatives offer a clear-cut solution. The ink remains red. Frustration begins to creep into everyone's voices. Bill Milton finally expresses it openly.

"Tell me your goals, give me absolute flexibility to achieve them, and then come up with a monitoring plan to make sure I'm achieving them."

"We've all got lives," he declared. "We don't need to run this ranch."

There. It had been said. Everyone seemed a bit stunned at first, and resistant. Then the unspoken realization was shared by everyone, that those words had brought a sense of relief and clarity to the discussion. Yes, it's true, everyone here is creative enough to have other things to do than to try to breathe life into this small sheep and cattle ranch on the Montana plains. And there are plenty of people who would tell them to get on with it. Ranching is a relic; the time has come to let it die. Perhaps we should all be helping it to die, or at least letting it.

But none of the people sitting at this table share that belief. They reject it with a passion that is the true wellspring of their creativity. They work to keep this ranch alive—not as history buffs trying to prop up some social and economic dinosaur, but because they believe it holds keys to the future, to all of our futures, which can be found nowhere else.

Bill and Dana Milton believe in America's agrarian myth because for them it's not a myth. They live it. They make their living off the land and live their lives as part of this rural community. Bill serves on the local PTA. Dana shuttles the kids to swim meets. Their daughter Moria wants to go to a good university somewhere in a big city, but when she graduates she wants to return

to a small town—because she believes the myth is not a myth, because she knows that diversity makes us more human. The Miltons see the rural agricultural community as a resource that the nation can little afford to let die. These are times, Bill reminded us that day, when the government is looking to partnerships with individuals and private firms to do everything from picking up trash along the highways to maintaining recreation sites in national forests and parks.

The only element keeping the government and rural land managers from coming together in a true partnership, Bill Milton maintains, is trust.

"The American public has to trust who's out there sustaining the land," he asserted. "It has to tell us what it wants: watersheds, recreation areas, wildlife, ecosystems, and biodiversity, too. Tell me your goals, give me absolute flexibility to achieve them, and then come up with a monitoring plan to make sure I'm achieving them. You're going to get me working to achieve your goals at no cost."

Trust a rancher? That's asking for more than most Americans would be willing to risk, but trust them with absolute flexibility on the land? That's ridiculous! And anyway, haven't we already made that mistake? Look at the overgrazed rangelands. Look at the devastated riparian areas. Look at the piles of dead predators. Livestock grazing has had its day of trust, and it blew it.

But things are different now, Milton points out. The way the Milton Ranch grazes its livestock today is as different from the way its predecessors did it in the nineteenth and early twentieth centuries (from the way some ranchers are still doing it) as ecosystem management is different from sustained yield. Instead of parking their animals on a piece of land until they gradually reduce it to a state of devastation where it's no longer economical to graze at all, the Miltons monitor their land and graze it according to what its biosigns tell them it will support. Instead of the traditional seat-of-the-pants approach of taking as much as they think they can safely take and then hoping there's enough left for next year, the Miltons start by keeping what they want to achieve in mind—an achievement that's as much a matter of the health of the land as it is a matter of cash flow and return on investment. What they do to achieve that goal has to pass tests on ecosystem health, endangered species, biological succession, and the type of energy they will use (solar power comes first.)

On a cold Montana winter day, ranch manager Rona Johnson checks in with her predator control crew. The Milton Ranch uses guard dogs to protect its sheep—an alternative to more lethal and less specific means.

And they don't do this on their own. They bring in other people to make sure that what they come up with isn't just wishful thinking born of tunnel vision.

To show me the results that process can bring, Bill Milton takes me on a tour of his land. The tour starts at an area that is currently being grazed and backtracks through the rotation system the Miltons use to keep their pastures steadily renewing themselves. First, we visit two pastures that are being grazed, one by sheep, the other by cattle. Both are close-cropped, but the one with sheep is more so. Sheep have become legendary for the devastation they brought to the West. Range wars were fought to keep them out. John Muir called them "meadow maggots" because of the devastation they brought to the high Sierra Nevada meadows that he loved so much. On the first pasture we visit, 2,000 sheep have been graz-ing for two weeks on a little over 300 acres. Here, it's easy to see what offended Muir so deeply. The ground nearest the sole water source, a small pond or "tank," is riven with tracks sliced and resliced into the soft mud. For several yards back from the water, the earth is stripped bare and worn

to rock-hard compaction. Across the rest of the pasture, the grass is sheared short, nearly to ground level.

From that pasture we move to one that was grazed two weeks before, then four weeks, then six. As we move back through the grazing rota-tion it becomes harder and harder to believe that these places ever looked like the one we visited first. At the last stop, which was grazed six weeks before, grass is growing to within a few feet of the water point (in this case a round metal trough). Since the trough is so much smaller than the pond in the first pasture, the impact of all those animals crowding in for a drink had to be several magnitudes more concentrated. Still, the grass is not only green but thick, almost right up to the drinker.

"Look at this land," Bill Milton exclaims, as I snap photographs. "Are you going to tell me that I'm hurting it?"

Some would, I tell him, because he cultivates the land; because he seeds some of it to species that didn't evolve here; because he manages it to produce a commodity—meat—from animals that also didn't evolve here.

"You know," he says, as we drive to a small lake that he has fenced from cattle and sheep to serve as a sanctuary for waterfowl and other wildlife, "what we're doing here is the cheapest, least ecologically damaging way to raise food on these lands." Less damaging than plowing them and seeding them to irrigated mono-cultures of wheat for whole-wheat bread or soybeans for tofu.

Bill Milton speaks of creating a grass culture on these lands—a culture that would take its inspiration from the culture of the Plains Indians, who once relied on the bison and antelope this land produced for sustenance. In his eyes, however, a contemporary grass culture would have to be a hybrid in order to exist—a hybrid for no other reason than that many pieces of the old puzzle are gone and would be difficult, if not impossible, to recreate. Gone, of course, are the great herds of bison and the vast unfenced spaces that gave them room to move, but those would be the easiest to restore. Gone also are the tribes of Plains Indians who hunted the bison and kept them on the move, torching the grass to move the animals and replenish their forage, slaughtering them in suffi-cient numbers to keep the herds within the capacity of their habitat.

Today, those old puzzle pieces have been replaced by a greater number of new ones that have never fit into the old slots: a new people and a new culture, a new land ethic and a new ethic with regard to animals and property. To fit with these new realities, Bill Milton's grass culture would use rotational grazing more than fire to revitalize the prairies. Rural communities rather than tribes would reap the benefits of that man-agement, sharing those benefits with the larger culture. "The bottom line," he says, "is having rural communities whose economic health is supported by healthy ecosystems."

When I called back a few months after my visit, to recheck my notes and see how things were going, much had changed.

"Did the financial team solve its problems?" I asked.

"We had another meeting," Dana told me. "We invited some of the neighbors over and made ourselves go through the planning model and wouldn't let ourselves be sidetracked. We threw everything on the table: selling the land, changing livestock, but when we figured every-thing in at the end we came out with a profit."

"We've got our operating costs lowered sig-nificantly relative to what they were," Bill added. "We're working on producing all our own feed. We're in the black, and things are going fine."

"What did you change?" I asked.

"We decided that our weak link was that I was still controlling the program," Bill admitted. "So, I stepped back. I'm on sabbatical now, and Rona and her husband Howard are managing the ranch. I went to a Zen retreat, then came back and worked with Rock Ringling of the Montana Land Reliance to put together a collaborative effort that includes point agriculturists and environmentalists, people from the Wilderness Society, the National Wildlife Federation, Audubon, and the Greater Yellowstone Coalition. We had a meeting to talk about how to improve Rangeland Reform '94, and we're planning another one in January."

Bill said his aim at putting together a state-wide team was to build deeper trust and under-standing. A lack of those elements, essential in any relationship, is what he believes creates the sort of vacuum into which the government feels obligated to move.

"If agriculturalists and rural states want to be victims, yelling and screaming is the best way to do it," he declared.

At last report, Bill's efforts at creating an alternative to acting like a victim were gaining attention. The governor of Montana was work-ing on establishing a center for conflict resolution for use by teams such as the one Bill was helping put together.

"It's a lot like Buddhism: very simple but very complex," Bill told me. "I don't think we'll ever arrive at a permanent result. I just want to keep the discussion going."

"Guard dogs don't herd sheep, as Border collies do; as puppies, they bond to the sheep as kin and protect them from predators, just as a wolf would protect a member of its pack."

"The most stable systems of grazing have been those
in which the experience, knowledge, and moral pressure
of a whole village guided the individual grazier."

DONALD WORSTER
from Under Western Skies

STIRRUP RANCH, COLORADO

Left to right: Rick Gossage, Paula Gossage, Kelly Coszalter, Louis Coszalter,
Dwayne Finch, Jane Wustrow, Jim Sazama, Burl Boren, John Valentine.

CHAPTER SEVEN

THE
STIRRUP RANCH

Rick Gossage, Paula Gossage, Louie Coszalter, Jim Sazama, and John Carochi

COLORADO

On Rambunctious Trout Streams,
Man-eating Grizzlies, and Subdivisions

A LINE OF BLACK CLOUDS rolling in from the west promised rain and lots of it as a Bureau of Land Management crew hurried to rip-rap down what was left of the riparian area along Badger Creek in south-central Colorado. Having seen what the rambunctious little stream could do, someone suggested that they park the bulldozer on higher ground than usual that night, just in case. Flash floods are common on this spring-fed trout stream that tumbles out of the mountains into the Arkansas River near Salida. Over the years, torrents of water laden with mud washed from abused range-lands and rocks ripped from the stream's boulder-strewn bed had stripped out Badger Creek's guts, so that now it raised more hell than trout.

The premonition was right. Night brought what the BLM calls a "storm of record," and 10,000 cubic feet per second of grit-filled, streambed-scouring slurry rampaged down Badger Creek. Two days later, when the crews were able to get back to their work site, they found the lower third of the creek's trout habitat blown out and the work that they had done obliterated. The bulldozer had managed to survive.

"That's when we got the point that it was going to take something more powerful than bulldozers and concrete to heal Badger Creek," said John Carochi, range conserva-tionist for the BLM's Canon City District.

But first, someone had to figure out what that something was. Carochi and a few others dedicated to finding a solution to the dilemma helped bring together a collection of eighteen government agencies and public interest groups to form the Badger Creek

Stirrup Ranch rangeland, where cattle numbers were increased six-fold as part of a carefully monitored management plan.

"If the developer had his way, the game trails… would be replaced by a network of bulldozer scars marked with street signs."

Watershed Project. In addition to the BLM, this planning and coordinating group included the U.S. Forest Service, the Soil Conservation Service, the USEPA, the Colorado Division of Wildlife, Trout Unlimited, and a few others who shared an interest in Badger Creek and the lands in its watershed. Regular meetings were held. Options were discussed. The creek kept flooding.

By the late 1980s, unable to come up with an alternative to what Carochi called "structures, structures, structures," even this approach had ground to a halt. "The project was ten years old and going nowhere," he admits. "The only ones who would show up at the meetings were government employees." Over the years, it had become clear that no amount of construction could anchor the stream's riparian habitat and no structure short of a dam that destroyed what it was designed to protect could tame its excesses. Now it had become clear that getting a bunch of agencies and interest groups together to talk about the problem wasn't going to get the job done either. As long as the land that fed Badger Creek shed runoff like a tin roof, the stream was going to continue decimating trout populations, blowing out stream improvements, and trying to eat bulldozers.

That's when Jim Sazama, BLM range conservationist, decided the secret to an effective solution might lie somewhere on the tin roof rather than along the streambed or in agency meeting rooms. The 135,000-acre watershed that feeds Badger Creek reaches its highest point on the summit of Black Mountain, a spruce-fir-clad extinct volcano that climbs to 11,654 feet. Wrapped around the

southern slopes of the mountain, the lands of the Stirrup Ranch serve as the keystone of the Badger Creek drainage. Sazama saw those lands as more than just the keystone to the watershed, he saw them as the key to the restoration project.

During the early years of the efforts to tame Badger Creek, the Stirrup Ranch had been owned by a corporation that had done little with the ranch except keep it on the real-estate market. In 1984, at about the same time that Jim Sazama was looking to the Stirrup for a solution to the stream's watershed problems, a developer made an offer to buy the ranch and was accepted. Subdivision, then, was to be the fate of this land of flower-carpeted meadows and clear mountain brooks. It would be no ordinary loss. In its long and colorful past, the Stirrup had been the primary haunt of Ol' Mose, the legendary last man-killing grizzly in Colorado. It had also been home to the man with the unlikely name of Whart (short for Wharton) Pigg, who bought the ranch so he could hound the giant bear into eternity. If the developer had his way, the game trails that Mose had once roamed would be replaced by a network of bulldozer scars marked with street signs. Maybe one of the signs would read "Old Mose Street," in honor of the great bear who was blamed for walking out of his den one spring day in 1883 and trying to eat the first thing he saw, which, unfortunately for both of them, happened to be Jack Radliff. Maybe there would even be a plaque telling western history buffs how Whart Pigg had been cheated out of his life's most consuming obsession when a houndsman he hired to help track Ol' Mose actually killed the bear that was "bigger than a wagon box." The story would surely be in the realtor's brochures.

"Own your own ranchette, your own piece of the American dream," the *Los Angeles Times* and *Wall Street Journal* ads might say. "Get back to nature in the home of the 'King of Grizzlies.'"

The implications for the Badger Creek Project, if the Stirrup was subdivided, were ominous. Instead of having to deal with one landowner, Carochi, Sazama, and the others trying to get the project back on track would have to deal with dozens, perhaps scores of them. If experience was an accurate indicator, a tin roof that was also a crazy quilt of small, privately owned tracts could be expected to produce more erosion rather than less. The Stirrup's fate was as good as signed, sealed, and delivered—except for one significant hitch: Subdivisions weren't such a great investment in 1984, and major league relief pitcher Rick (Goose) Gossage, who was out to fulfill a lifelong ambition of owning a ranch, had also made on offer on the ranch. Gossage had earned his fame playing in the World Series for the New York Yankees in the 1970s. Some say he's headed for the Hall of Fame, but in his heart he has remained a country boy. Raised in Colorado Springs, only a couple of hours from the ranch, Rick has always loved the mountain country with a love nurtured in him by his father. "My dad grabbed every spare moment to take us fishing or hunting for arrowheads up in those mountains,"

"In Colorado, every hour of every day, four acres of wildlife habitat is gobbled up by development."

Native grasses and wildflowers grow profusely in the Stirrup's barnyard, indicating that even land that has suffered prolonged periods of the most severe overgrazing and the heaviest soil compaction can respond to alternating periods of disturbance and rest.

he recalls. When Rick found out the Stirrup was for sale, he had just signed a lucrative contract with the San Diego Padres. "I felt the best place I could put that money was in the land," he said.

Gossage made his offer and waited. When he finally did get a response, it was to tell him that the developer's bid had been accepted instead of his. He was crushed. But not too long after, he got a second call. This one was to inform him that the owners had reconsidered and opted to take less money for what they felt was a better risk. The ranch was his.

Freed from the tracks with the train bearing down at full speed, the Stirrup was saved, at least for the time being, from what may be the greatest villain that has ever roamed the West. These garbage-spewing, habitat-gobbling, air-polluting collections of dusty roads, incongruous-looking houses, and backyard feedlots, full of yapping dogs and off-road-vehicle-crazy kids, that we call subdivisions threaten a final solution to the question of what to do with all those wildlands and the charismatic creatures that haunt them. Though mountain lions have survived all the traps, poisons, lead, and hounds a century's worth of ranching has thrown at them, they're not surviving subdivisions. Though ranchers and cows are considered instrumental in banishing natural fire from the West, wildfires and houses too far from a fire department make even worse bedfellows.

Subdivisions are considered such a serious threat by some that ranchers use them as a kind of bogeyman when the debate over public lands grazing gets really hot. "If you chase us off, we'll sell our private lands to the subdividers," they hoot.

Those who strive to make the West "cattle free" say this trump card is nothing more than an empty threat; that either the private lands associated with more than 31,000 grazing allotments will be bought up by the government, or they'll never sell because they're more than the market will ever bear. Neither of those arguments seem to fit the contemporary West. The government can't afford to manage the land it already owns (even Grand Canyon National Park has had to take hat in hand and go begging to private corporations), and the real-estate market in the West seems to have a nearly insatiable appetite when it comes to land, no matter how remote or expensive. In Colorado, every hour of every day, four acres of wildlife habitat is gobbled up by development. "Ranches wanted, top prices paid" is an ad that

"Subdivision is the greatest threat to both western agricultural land and wildlife habitat."

you can read in virtually any real-estate multiple-listing flyer in the West.

Most of the old ranches around the Stirrup are gone. In spite of ranchers' protestations that they stand as guardians of the land—the simple fact remains that when the price gets high enough, they sell out. Dwayne Finch, Colorado Division of Wildlife officer for the land where the Stirrup is located, says, "All the old ranches have sold around here; the kids have moved out, they're not interested in the lifestyle. A lot of 'em have sold to subdivisions. Problems?" he reflects, "I can think of a thousand problems they cause: poaching, abandoned shacks, junked cars, more roads, power lines, erosion...."

Faced with that as an alternative, many environmentalists are striving to find a way to keep ranchers on the land, if it's possible to do that and have healthy rangelands as well. "Nothing makes more sense than for ranchers and environmentalists to work together," claims Bill Branan, manager of a National Audubon Society nature preserve in Arizona. "Subdivision is the greatest threat to both western agricultural land and wildlife habitat."

But some environmentalists think anything is better than a cattle ranch, and a number have said as much. "I don't want you to quote me as saying I want there to be subdivisions all over the West, but I would say that a subdivision has less environmental impact than a cattle ranch," said one lawyer who has won lawsuits against ranchers. When I hear such statements, the only way I can make sense of them is to figure that they're confrontational blustering, calculated to attract contributions from other confrontationalists. I live in a subdivision. Its main claim to diversity is that its residents use more than one brand of weed killer. I've seen ranches like the ones in this book. I've seen more wildlife on them in one day than I'll see the rest of my life in my yard.

With the Stirrup's demise averted, at least for the near future, Jim Sazama was ready to proceed on the premise that signing up a high-profile cooperator was the best way to breathe life back into the Badger Creek Project. With high hopes, he approached the new owners of the Stirrup, but instead of coming to them with a proposal, he came with a question: How could the ranch and the agencies and groups involved in the project work together to increase the amount of plants covering the ground on the ranch and slow down the water flowing off it? The floods coming

down Badger Creek told the story that the Stirrup was in bad shape, but they didn't tell exactly why. The ranch really didn't look that bad.

Since it was settled in 1870, the Stirrup has always had a history of relatively light use. Mostly above 9,000 feet, with short growing seasons, cool summers, and unpredictable moisture, the ranch has always had a reputation as "rough sleddin' for a cow outfit." The previous owner had leased portions of it for light grazing, but as Sazama describes, "Parts of the ranch would go for years with no use at all." As a result, much of the Stirrup was covered with the tall waving grass most of us think of when we imagine healthy rangelands.

When Rick bought the ranch, he and his ranch manager, Louie Coszalter, kept the grazing pressure light and waited for the land to improve. It responded by remaining pretty much the same. If anything it seemed to be looking less, rather than more, vigorous. One evening, after Sazama and Coszalter returned from hearing a talk on innovations in range management, they walked out into the grass behind the ranch house and looked at it with new eyes. What they saw surprised them. "I think that's the first time I looked at those plants and thought, 'They're big and tall, alright, but they're dying,'" said Sazama. Most of the leaves of the Idaho fescue and slimstem muehly plants he was looking at were brown and dry. That meant they were at least a year old, maybe two. Some were gray and nearly black—that meant they were even older. Most of the plants showed a few green leaves among the brown, but a significant number showed no sign of life. They were apparently dead. When touched, their leaves shattered as if they were nothing but ashes. Mummified by the drying sun and wind, oxidized until they really were little more that ashes, these standing corpses were devoid of nutrition. That is why they were still standing. Neither cattle nor elk would eat them. Instead of being overgrazed, the area was grazed very little. In spite of that, more plants were dying than were being replaced by seedlings. The "tin roof" was expanding.

Their eyes opened, Jim and Louie now began to realize that their strategy of being gentle on the ranch wasn't having the desired effect. With that in mind, they decided to take a class in Holistic Resource Management. John Carochi, Rick, and Paula Gossage, Rick's sister, signed up later. "We needed a new way of looking at the ranch,"

Carochi explains, "and HRM was the only new way around."

At the time that she took the class, Paula was the ranch's office manager. A crystal-wearing, mountain-biking, soft-spoken woman in her early forties who had grown up in small towns that had not yet lost their rural feel, she ended up going to the HRM class, though she said she didn't see any reason to at the time. Afterwards, she came back with a different commitment to the ranch—and with a new outlook on life. "That class changed my whole life," she says now. "All of a sudden, what I was seeing on the ranch made sense." In addition to teaching her about grazers and grass, the part of the class that dealt with planning, goal setting, and teamwork inspired Paula. She became the ranch's projects coordinator and meeting facilitator. In that capacity, she began serving as an advocate for the Stirrup's involvement in the Badger Creek Project and for using the ranch as an educational tool to help sign up others for the project. She began facilitating team meetings.

The class also changed Rick and Louie's idea of how the ranch should be run, and the BLM's, too. Equipped with a new point of view, they started working together to come up with a new management plan intended to finally put some "hair" on the tin roof. Because Louie hadn't grown up in a ranch family, accumulating a history of bad experiences with agencies, Sazama found that brainstorming new ideas with him was easier than it might have been if he were fourth-generation ranch stock. "Louie wasn't ground down and worn away. We could challenge one another. We were in a nonoffending mode with one another," explains Sazama.

After assessing the ranch's carrying capacity using the HRM model, Louie described the numbers he and Sazama came up with as "phenomenal." If the data were accurate and the planned actions based on them passed all the other tests posed by the planning model, theory had it that they could introduce those increased numbers without overgrazing the land. The payoff, if all this worked, would be more diverse, healthier grasses for both domestic and wild animals, and less bare dirt to send floodwater raging down Badger Creek. They decided to give it a try.

Before the Stirrup Ranch could fully implement the plan, steps had to be taken to build the structures to accommodate the new system and to protect various aspects of the ranch's ecosystem

"The payoff, if all this worked, would be more diverse, healthier grasses for both domestic and wild animals, and less bare dirt to send floodwater raging down Badger Creek."

from suffering inordinately from the increased densities it would bring. "We didn't have enough water in the right places to distribute the animals the way they had to be distributed," Coszalter relates. Alternate water sources would take the pressure off riparian areas where larger numbers of livestock would have a tendency to congregate. More water sources would also make it possible to move the cattle frequently enough to keep them from overgrazing.

Another problem that had to be solved, according to Coszalter, was to come up with a way to keep the animals "bunched up," so the impact that they would have on the land would be similar to that of naturally herding grazers. Coszalter didn't want to use the solution most ranchers use to get this job done, an extensive network of fences, so he decided to use cowboys to herd the animals instead. "It worked okay during the day," explains Coszalter, "but we'd go home at night, and when we came back in the morning, they'd be all over the place." Finally, Coszalter and Gossage decided to give up on herding and build the fences.

The ranch paid for and installed eighteen miles of electric fence that was low and unobtrusive but effective. Other elements of the required infrastructure were paid for on a cost-share basis, with the ranch putting up part of the money and the rest coming from a combination of federal sources, including the U.S. Soil Conservation Service, the local Grazing Advisory Board, and a grant from the Environmental Protection Agency (EPA) under Title 319—a program targeted at decreasing nonpoint water pollution on rural lands. John Valentine, who worked for the Sangre de Cristo Resource Conservation and Development District at the time, was instrumental in getting that grant. When it all finally came together, says Valentine, the project represented the first time in the nation's history that so many government agencies and private entities had cooperated to tackle so large a rural watershed project.

With the infrastructure in place on the ranch, Coszalter began cranking up the number of animals under Sazama's watchful eye. Starting from a base of about 300 head on the west side of the ranch, the Badger Creek side, Sazama relates, "The first year we decided to go a little heavier and went to 600. When we went to look, they were so spread out we couldn't find them." The next year, Sazama reports, they increased the

Prairie flax provides a touch of diversity and color on the scenic Stirrup Ranch.

number of animals to 1,000 head, with much the same results. A couple of years later they went to 1,800 head and still the land showed no apparent ill effects. In five short years, the number of animals on the west side of the ranch was increased by a multiple of six, and the only change the BLM's monitoring program could detect was an increase in ground litter.

While this was happening on the west side of the ranch, Louie and Rick began increasing numbers on the east side as well, and rotating the animals in similar fashion. Eventually, the total number of cattle on the ranch was increased from 600 head to 2,800.

If all those animals didn't hurt the land, did they help it? Sazama said that the fact that they were able to increase the number of cattle so significantly without causing any observable damage served for him as an indication that they were on the right track. Though their monitoring methods showed little change, positive or negative, walking on the land told a different story.

Before-and-after photographs indicated enough improvement that Sazama granted the ranch the increase in permit numbers.

The story the land told to Paula Gossage was one of affirmation. "The quality of the grasses was better," she said. "There was less of that dead stuff, and the new grass coming in was greener. And the riparian areas—the change there was wonderful." The riparian areas, Paula said, were literally coming back from the dead. Willows were once again growing along the streambanks, and bare places where the animals had once congregated were filling in with grass.

John Carochi was so impressed with the results that he began lining up field trips to take private citizens and other agency personnel on tours of the ranch. "We've shown Sierra Club members, Earth First!ers, and other ranchers that you can make a difference with livestock, and nobody who has come to see has disputed it," enthuses Carochi. "When someone says this is all bullshit, we can go out and look at the Stirrup."

But the most enthusiastic review for the program didn't come from people, it came from elk—the elk that use the ranch as winter range. As recently as 1967, elk were being imported to the ranch from Yellowstone National Park. Before that, Whart Pigg tried ranching elk on the Stirrup in the early 1900s, and many of those animals escaped. Before the change in management, there were elk on the ranch, but not lots of them. After the change, their numbers increased dramatically.

Wildlife manager Dwayne Finch has lived in this area for twenty-one years and knew the Stirrup before the Gossages bought it. "All my life I thought elk lived in the quakies and belly-deep grass," he mused, as we sat around the kitchen table in the ranch headquarters. "But you know where they winter now? In the short grass on this ranch. Why? Probably ought to ask an elk, but I think it's because of the diversity and the quality of the food. That probably comes from the way this place is being managed."

According to Finch, 1,500 elk now winter on the Stirrup. They come to eat the tender, fresh growth that resprouts after Louie moves the cattle out of the high pastures. As part of the ranch's management plan, Coszalter makes sure he does that early enough in the season for the grass to regenerate before the elk move in.

Though ranches are made up of elements that are enduring if not eternal—mountains, ecosystems, streams, seasons, government agencies—ranch managers are much more fragile. A lot of things can go wrong with anything that involves human lives and passions. It's ironic that events as far removed from dirt and grass as personal relationships and big-league baseball can put watersheds in jeopardy or cause less grass to grow; however, in 1991, the foibles of human relations almost took down the keystone to the Badger Creek Project. First, Louie and Paula got a divorce. Then, there were financial problems, complicated by the fact that the ranch finances were not differentiated from Rick's baseball income. All of a sudden, the Stirrup was back on the market. Management innovations were put on hold. The cattle were sold, and the ranch went back to leasing its private rangelands to other grazers. Paula moved to town, although she remained active in the affairs of the ranch.

Through all of this, the ranch never went back to grazing the land in the old way. "We grazed someone else's cattle, but we took care of them ourselves," Louie said. "We didn't let them graze just anyplace or any way they wanted to."

For a while, it looked as though the ranch would be sold to a sportsmen's club to be maintained as a hunting preserve, but the deal never did go through. Then, as time went on and the ranch didn't sell, things began to come back to a state of order and equilibrium. Louie remarried. Financial planning and management practices were implemented that resulted in greater financial stability. In the end, after having lost only two years in the recovery program, Rick, Paula, and Louie have become enthusiastic about the ranch again. Inspired by their enthusiasm, they are talking about buying some bison to add another element of diversity to their management. Rick is looking forward to spending more time on the ranch after he retires from baseball, if not this year, then soon.

As this is written, Paula is in the process of moving back to the ranch, where she hopes to build on the momentum started by Carochi and Sazama's tours and work on her dream of turning the Stirrup into an environmental education center and retreat. She's even thinking about doing some outside consulting as a team-building facilitator.

"I've thought about the ranch a lot in the last few years—that I've learned a lot from it," Paula told me as we sat on a high point overlooking much of the Badger Creek watershed and the lands of the Stirrup. "Even if we end up selling, I feel good about what we've done here. I feel that we've started something that can now stand on its own. We've helped change the way Jim Sazama and John Carochi and the other people on our team look at the land, and we've helped create enough enthusiasm among the other ranchers in the watershed that they're doing things that I think go beyond what we have done. They're the ones you ought to be writing about now."

"We've shown Sierra Club members, Earth First!ers, and other ranchers that you can make a difference with livestock, and nobody who has come to see has disputed it."

"We would do better to approach the cattle frontiers as humble humanists than as confident scientists, if seeking the truth is our goal."

TERRY G. JORDAN
from North American Cattle Ranching Frontiers;
Origins, Diffusion, and Differentiation

SIEBEN RANCH, MONTANA

Left to right: Neill Windecker, Mesia Nyman, Chase Hibbard, Tom Clifford, Graham Taylor, Mark Ahner.

Not pictured: Ann Hibbard, Mike Aderholt, Bill Long, Dave Cole, Ron and Debbie Ries.

CHAPTER EIGHT

THE
SIEBEN RANCH

Chase and Ann Hibbard

MONTANA

*On Cattle and Elk, Win-Win, A Devil's
Brew, and Fresh Forage*

THINKING OF ELK as a problem doesn't sit well with Chase Hibbard, owner of the Sieben Ranch northeast of Helena in west-central Montana. "I love to see them," he declares. And there are plenty of elk to see on the Sieben's more than 60,000 acres at the foot of the Big Belt Mountains. The ranch is located adjacent to the Beartooth Wildlife Management Area (BTWMA). Locally referred to as the "game range," the Beartooth is home to more than 1,500 elk. From 500 to 1,000 more elk live on the ranches that border it, including the Sieben.

As we sat in Hibbard's log cabin ranch residence and looked out across bright green irrigated meadows to pastures of native grasses, we couldn't see any elk, but we could see antelope. On my way to the ranch, I had driven past miles of grass and wildflowers and I didn't see any elk, but I did see a couple of deer standing on top of a haystack in a barnyard and several more eating fresh barley sprouts in one of the crop fields. "The elk are up in the high country right now," Hibbard told me, as he opened a bottle of Elk Cove Reserve pinot noir. "They're enjoying the cool air and eating the grass that grows during the extremely short growing season up there, but they'll come back, after the snow flies, and they won't leave till spring."

Every winter, elk drift down out of the Big Belts with the snow, as they have done since long before Chase's great grandfather, Henry Sieben, bought the ranch in 1907. The BTWMA used to be a ranch, too—the ranch that belonged to Bill Milton's family, the one he saved from subdivision by convincing the Nature Conservancy to buy it

65

in 1972. The Conservancy sold the ranch to the State of Montana, which agreed to designate it a wildlife preserve and abide by several other restrictions the Conservancy attached to the deed. In a state where hunting is still king, the game range has been managed almost exclusively as big-game habitat since then.

When the Conservancy bought the Milton Ranch it was home to only about 300 elk. Since livestock have been removed, the elk herd has tripled. That fact is deceiving, though, says Hibbard. Those elk don't live only on the game range, and their increase has been at least partly caused by improved conditions on adjacent ranch-lands. When they drift down out of the Big Belt Mountains to add their numbers to the 500 to 1,000 other elk that live on those mostly private lands, they end up in irrigated hay meadows, in fields of barley and oats, and in pastures that local ranchers have planned to use as spring feed for winter-hungry cattle. Then they become a problem—a problem that has been getting bigger every year. "You can't keep them out," Hibbard grumbles. "We've tried cracker guns and propane exploders, but oats especially are just like candy to them."

This land has always been productive land. That's what brought Henry Sieben here more than a century ago. Sieben first came to Montana

with his brother in the late 1860s, and together they started freighting supplies in wagons drawn by teams of oxen from Fort Benton, on the Missouri, and Corrine, on the railhead in Utah, to outlying mining camps. At one point, Sieben bought a bunch of used-up oxen and turned them loose in the fertile Montana grasslands to graze the green growth of summer. At the end of the season, he found that they had gained enough weight to earn him a handsome profit. Always alert for an opportunity, Sieben began running cattle on the open range between Helena and Great Falls and noticed at the end of summer that his stock always ended up at the head of Hound Creek, at a place called the Hole in the Wall. That was enough to make the business-wise old pioneer vow that if he ever owned a ranch, that is where it would be, where the grass made the cattle fat. When the Cannon Sheep and Cattle Company put the Hole in the Wall up for sale in 1907, Henry Sieben kept his vow.

Since then, four generations of Hibbard's family have ranched this country, raising cattle and sheep. And the elk have never really been a problem, until now.

Big numbers of elk are becoming increasingly common throughout the West. In many places, herds have grown to sizes greater than anyone can remember. Pushed by growing base populations that are outstripping the capacity of their habitat, the animals are moving into areas where no one can remember ever seeing them before. Although we think of elk in terms of trees and mountains, they are predominantly a grassland animal, and it appears they are ready to reclaim their position in the grasslands of the West.

As elk numbers increase, people are reacting in different ways. Some like it, some don't—at least not all of the time. For many of us, that's hard to understand. Our image of natural beauty usually has an animal in it, one that looks like a deer—and elk are big deer. Generally, if people live where there are elk they enjoy seeing them, hearing them, and knowing they are there, but even the best of friends can wear out a welcome.

The way to get to not like elk is to have some sort of conflict with them. Since a lot of ranchers live in elk habitat and make a living off the same grass that elk need for survival, you might say ranchers and elk have the makings for a classic conflict of interests. Sometimes that conflict can get serious.

In Arizona, in 1987, a rancher was convicted of shooting at least eleven elk, and maybe as many as twenty, and then burning and burying their carcasses to cover up what he had done. He said he did it because he was afraid that the increased impact the animals were having on his forest service grazing allotment, along with the impact his cattle were having, would cause the forest service to cut the number of cows he could graze there. So, he started shooting elk.

In 1985, a rancher who doesn't live far from the Sieben Ranch sent the Montana Department of Fish Wildlife and Parks an invoice for $108,915 for the grass consumed and damages caused on his private land by elk and other wild ungulates. He said he considered it pasture rental. The rancher claimed the elk ate so much of the grass already weakened by drought on his private property that he was faced with cutting his cow herd by 20 percent.

Back in Arizona, efforts to slow the growth of an elk herd the state game and fish department claimed was contributing to serious habitat deterioration turned into a national issue when a Maryland group succeeded in getting a special hunt stopped with a court injunction. The hunt would have reduced the number of cow elk in an area where drought, excessive cattle use, and an elk herd that had partly changed from migratory to resident had stripped the land, in some places, to bare dirt. Ranchers had been calling for such a hunt for some time, saying the elk had grown in numbers to the point where it made managing the land impossible. Cutting their cattle numbers hadn't helped. Even moving cattle completely off some areas hadn't helped. They said that efforts to raise more grass to mend devastated rangelands just attracted, even grew, more elk.

Hunters, wildlife groups, and animal rights activists opposed the hunt, painting it as a case of the government caving in to greedy ranchers who were just trying to get more grass for their cows. Whatever grass grows on public lands belongs to elk first, they protested, and if there's any left over, maybe some of it could go to livestock, but only after every last wild creature has had all it wanted.

The Arizona Game and Fish Department countered that their concerns went beyond finding someone to blame. "The opponents of the hunt didn't challenge our justification for it. We were

"Generally, if people live where there are elk they enjoy seeing them, hearing them, and knowing they are there, but even the best of friends can wear out a welcome."

Chase Hibbard stands next to another indicator of the beneficial effect animals can have on the land: a clump of tall grass growing on an anthill.

ready to defend that," explained Tom Britt, Region 2 director for the AGFD. "They got us on a technicality." The hunt was canceled at 4:50 PM for insufficient notice to allow public comment on its first day. By 5:30 PM, Britt had his officers flying search grids over the area. Vacations were canceled, and anyone who could be spared became part of the search crew. For two days and three nights, AGFD personnel flew, drove, and walked the hunt area telling permit holders that the hunt had been canceled. In spite of those efforts, voluntary survey forms showed that at least 144 animals of an initial goal of 200 were killed.

Before the situation on the lands surrounding the Beartooth even hinted at escalating to that same state of contentiousness and absurdity, Chase Hibbard and a few of his neighbors decided they'd better do something about it. So, they got together one Sunday with Bill Milton, whose family owned the Beartooth when it was a ranch, and with a few others to kick around some ideas about how they might call off the war before it passed the point of no return. The only way to do that, they decided, was to bring all the players together to "dream a larger vision of mutually shared management."

"We felt there was something inherently wrong with the process as it was," Hibbard explains. "The only way to play was to complain the loudest and maybe someone would listen."

Hibbard, Milton, and the others then pulled together a larger group to see if they were willing to work toward the same dream. The second meeting included the district forester from the Helena National Forest, the regional supervisor for the Montana Department of Fish, Wildlife, and Parks, representatives of the BLM, the State Lands Department, and sportsman's groups that used the Beartooth. All the ranchers in the neighboring area attended, too. Together they decided to form a group and call it the Devil's Kitchen Management Team. The team would be based upon the principle of Win-Win or No-Deal. Not on compromise. Not on consensus. Everybody had to agree that they were coming out a winner or they wouldn't agree at all. Meetings would be facilitated by the Montana Land Reliance, which Hibbard helped form with Bill Milton.

The 250,000-acre area, for which the group was trying to achieve cooperation before conflict, includes four large ranches and several small ones.

It also includes a forest service wilderness area (the Gates of the Mountains Wilderness Area) and the state-owned Beartooth Game Range. It is bounded on the north by the Missouri River and on the south by the Smith River, one of the top trout streams in the nation.

The team chose the name Devil's Kitchen from a rocky labyrinth within its area of concern. This jumble of cliffs, pinnacles, and passageways provides a puzzle to those who encounter it, just as the issue they were addressing would provide a puzzle to the group. There was another reason they liked the name. They liked the idea of being able to look at themselves and what they were doing in a lighthearted way, as if they were stirring up some mysterious concoction full of all sorts of incongruous ingredients that everyone was suspicious of—a devil's brew of bat wings and eyes of a newt. Maybe the image would make them smile and help pull them through the times when they were at each other's throats. They knew those times would come.

For four years, the team worked to put together a proposal to take to the Montana Fish and Game Commission—one that would make everyone a winner, including the elk and the ecosystem. During that time, Chase recalls, "Every human emotion came up: disgust, anger, elation"; but the group persevered, forging a base of trust and acceptance from the heat of confrontation. As Hibbard tells it, "In most groups everyone shows up with their own agenda. After working together for a couple of years, we began to see past our own narrow interests. We began to see that whatever any of us did affected all the rest of us." In short, the team began to think like an ecosystem.

In 1992, the Devil's Kitchen Management Team approached the Montana Fish and Wildlife Commission with a proposed set of elk-hunting rules for their area that would have achieved the goals of all parties involved. They were turned down. In 1993, they came back with a slightly less-ambitious proposal and were successful. "We not only got the commission to adopt our proposal," states Hibbard proudly, "but they told us that whatever our group decides in the future will be what happens with hunting seasons in our area. That's never been done before anywhere in the West, so far as I know."

The commission decided that, as long as the Devil's Kitchen Team continued to work with all

"After working together for a couple of years, we began to see past our own narrow interests. We began to see that whatever any of us did affected all the rest of us."

the interested players, the proposals it brought to the commission would be given the same weight as recommendations of the department's own staff. Actually, clarifies Mike Aderholt, wildlife manager for the region that includes the Devil's Kitchen area, the team's proposal came to the commission a step ahead of a department proposal. "They'd already hammered out all the compromises in the team meetings," says Aderholt. "It just sailed through."

Members of the team were elated…and a bit surprised. "There was some controversial stuff in that package," remembers Mesia Nyman, who is assistant district ranger for the Helena District of the Helena National Forest and a team member. "The proposal limited hunting on public lands as well as on private lands. The organized groups would never have gone along with that if their representatives had not been part of the process."

The Devil's Kitchen Team's proposal would keep the entire hunting season of six to seven weeks open for cow elk and limit the kill of bulls on the area to eighty, at least partially by means of a voluntary quota. The intent of these measures was to achieve the goals of keeping the size of the elk herd in harmony with the carrying capacity of the land, increase the age diversity and percentage of bulls in the herd, and satisfy the public's demand for more access to more land.

In spite of its success, or more likely because of it, the Devil's Kitchen Management Team has critics as well as admirers. "We've been told that [by accepting this proposal] the department is abdicating its responsibility, and that we're contributing to the privatization of public lands," reveals Mike Aderholt. As a result of the agreement, though, he points out that there were more hunting opportunities and more opportunities for access, not less.

For members of the team, the group's success goes beyond the fact that it has changed wildlife management in Montana, perhaps forever. Chase Hibbard describes the personal growth he has experienced as a participant in this process as one of the most fascinating experiences of his life. That's saying something. Chase has been around. As a kid, he spent school years in Helena and summers on one of the most beautiful ranches in America—a rich city kid and a Montana cowboy all rolled into one. After getting a degree in political science at Amherst, he spent six years in commercial banking in San Francisco. That's where he

and his wife, Ann, were in 1976 the day word came that his father had been killed while flying his private plane over the ranch. "One day I was a banker, and the next day I was a rancher," Hibbard states simply. "I had to learn fast."

Since his return to Montana and the ranch, Hibbard's life has been even more diversified. He is deeply involved in both rural and urban communities. He represents an urban district in Helena in the Montana State Legislature and is active in a number of agricultural associations, including the Montana Stockgrowers and the Montana Woolgrowers. (The Sieben Ranch has the last remaining commercial band of sheep in the Devil's Kitchen area.) As an expression of his concern for the perpetuation of rural culture, he served several years on the board of the Montana Land Reliance, working to secure conservation easements on ranches and farmland. In 1988, he narrowly missed winning the Republican primary race for lieutenant governor.

Through all this, Chase has managed to be a successful rancher, and one that takes care of his land as well as his business. The ranch has always been in relatively good shape, Chase explains, because it's never been overgrazed, due to limits on essentials such as winter feed and areas for calving and lambing. About ten years ago, however, Chase became concerned about the condition of lower-elevation riparian areas in his summer pastures. They were grazed especially hard both late and early in the season, when cold weather and snow kept the animals out of higher country. As a result they were suffering, and they showed it. "They'd been getting hammered for years," Chase recalls.

Looking for a way to remedy the problem, Chase toured a ranch that used a management system set up by rest-rotation grazing guru Gus Hormay. Hormay has been setting up ranch management systems since the 1940s. Hibbard was so impressed by what he saw that he invited Hormay to put together a plan for the Sieben. Since Hormay had already retired, he refused at first, but changed his mind a couple of weeks later and called back to say he would do it. Hormay put together a management system for the ranch that incorporated a version of the three-year rest-rotation cycle for which he has become famous. Under this system, each pasture would be rested year-round for one year, then grazed year-round for another. For the third year, pastures would be

"Chase Hibbard describes the personal growth he has experienced as a participant in this process as one of the most fascinating experiences of his life."

rested through the growing season and grazed only after most seeds had ripened. Hormay's system is less intense than the rotation systems used by Rukin Jelks and some of the other ranchers in this book, but it is quite effective in the cooler, more moist conditions found in northern Montana. Hibbard is still fine-tuning the system, but results are already impressive, and he's eager to show them off.

Ironically, it is in areas that Chase says he once considered to be in the worst shape that he now gets to show off the most dramatic success. In these holding pastures and hangout places around outlying barns and water, grasses and wildflowers now reach nearly to the window of Hibbard's pickup. Later, on another tour, I rode my mountain bike and took a photograph of my wheeled version of "Old Paint" nearly hidden in this tall grass. Near an old sheep barn, on land that had been grazed as hard as a piece of land could be grazed for more than a century, western wheat grass reached to my nose.

At one point on our tour, Chase climbed out of the truck and dropped down on his hands and knees to show us one of the photo-monitoring sites he had set up earlier with John Siddoway of the U.S. Soil Conservation Service. Pushing aside the grass, he said, "Look at all this bare ground." The distance between live perennial plants that he was referring to could have been measured in inches or even fractions of an inch. I told him that I'd been to plenty of ranches where those distances are measured in feet, in some cases, yards.

Farther on, we visited a second monitoring site that was on the Beartooth Game Range. At the time of my visit, it had been three years since the Sieben Ranch and the State of Montana had reached an agreement to permit Hibbard's cattle to graze the range. That was the first time livestock had been on the game range legally since its designation as a wildlife reserve. "Why do we graze the game range?" Chase repeated my question after I asked it. "We don't need the pasture, and we're not increasing numbers, but we're still

A sheepherder and his guard dogs head out to catch up with their band of sheep on the Sieben Ranch in western Montana. The Sieben also uses llamas to guard sheep, entrusting their flock of prize rams and ewes to these animals that are aggressive enough to ward off coyotes.

paying rent. We're doing it for selfish reasons. It helps us avoid a bottleneck in our rest-rotation system, and we want to keep those elk at home, so we freshen up their feed a bit."

Hibbard was referring to a phenomenon that has been observed in grassland ecosystems around the world. On the Serengeti, gazelle follow wildebeest, eating the fresh regrowth that springs up after almost two million mouths have eaten and passed. Elk are meticulous about eating only green grass, if at all possible. They'll pass up hundreds of tall, dry, bleached-out grass plants on the way to one small green one. Extensive stands of grass on the game range were in an advanced state of stagnation, and therefore of little use to elk, when fire burned 32,000 acres of the range in 1990. Though we tend to think of fire as a killer of plants, in this case it actually saved many of those grasses from smothering under their own accumulated waste. The fire had its negative effects, too. It burned nearly 90 percent of the thermal and hiding cover used by the elk. Since the fire, grazing by Hibbard's livestock has served to keep the stagnation from recurring, while also avoiding the re-creation of the hazard that fire poses to elk cover and surrounding forests and ranches. And it has given the elk more green biomass to eat at home, so they spend less time in ranchers' fields.

The game range fire had another positive impact on the environment of the Bigtooth Game Range: it provided an opportunity for the Devil's Kitchen Team to test its mettle and to prove its value to the area, raising its own confidence level in the process. "We were faced with a real crisis right after that fire," Chase recalls. "There were 1,500 elk who had literally just been burned out of house and home." Because the landowners in the area were used to working together, Hibbard says, they could get together in a hurry and plan how they were going to handle the crisis. Hibbard ended up feeding 600 extra elk that year, at least until the grass regrew. Because everyone agreed to share part of the burden, no one was hammered beyond what they could provide.

"We had that problem solved in two-and-a-half hours," Hibbard remembers. "Without the team we'd have had a disaster. I think you can call that a success."

Success like that is bound to draw attention, and it has. The promise the collaborative process, as used by the Devil's Kitchen Team, holds for other hard-to-solve disputes with multiple players has attracted attention beyond the grasslands along Hound Creek.

Team member Mesia Nyman points out that the group and its process are already being used as a model in and around Montana and beyond. The team has been designated one of five National Common Ground Demonstration Areas by a committee set up at a 1991 symposium called "Livestock/Big Game: Seeking Common Ground on Western Rangelands." The symposium was sponsored by more than a dozen agencies and groups representing land livestock and wildlife interests. Just how easy it will be to duplicate the Devil's Kitchen success story is a matter yet to be discovered. "Success certainly isn't guaranteed," warns Nyman. "So much is dependent on the group members' abilities to see beyond their own single focus." Nyman admits she has been impressed by how well members of the Devil's Kitchen group have been able to do this. And she's also been amazed by the energy and staying power the group has shown. "We must have hit the right combination," she concludes.

At a time when problem solvers are in more danger of being oversold than underchallenged, Hibbard cautions that any broadening of the group's focus must evolve via the teamwork process. He does, however, name other areas of controversy that have been discussed as opportunities for building on the success achieved on the elk issue. Among the challenges waiting to be tackled are logging policy, subdivision of rural lands, and management of other species, such as deer and trout. But the only way the group will choose to apply itself to any one of those issues is by the same win-win or no-deal process that brought it success in one of the hottest natural resource issues in the contemporary West.

"The promise the collaborative process…holds for other hard-to-solve disputes with multiple players has attracted attention beyond the grasslands along Hound Creek."

"Humans and Nature had evolved together to
form a system that sustained a rich diversity of species,
something that stirred poets and even Darwin…"

JAMES LOVELOCK
from The Ages of Gaia

PITCHFORK RANCH, WYOMING

Left to right: Joe Hicks, Bob Edgar, Lili Turnell, Jack Turnell, Ray Mills, Jerry Longobardi.

Not pictured: Joe Thomas, Dr. Quentin Skinner, Mike Smith.

CHAPTER NINE

THE
PITCHFORK RANCH

Jack and Lili Turnell

WYOMING

On Endangered Species, Recovery Teams, Riparian
Areas, and Being Under the Gun

IN 1981, ONE OF THE MOST ENDANGERED ANIMALS on earth, a black–footed ferret, turned up dead in Jack Turnell's neighbor's yard near Meeteetse, Wyoming. When that animal was traced to Turnell's Pitchfork Ranch, it changed his life forever.

"It came right in the yard and the dog killed it," Turnell's neighbor, John Hogg, recalls. Hogg said he heard his dog barking in the night and found the odd-looking creature the next morning. Because he had never seen one before, he didn't know what it was. "It looked like a mink," Hogg told me, "but I knew it wasn't a mink." Hogg showed the odd-looking animal to his wife, Lucille, and then he threw it over the fence to get rid of it, as he might an old scrap of hide the dog wouldn't eat.

After reflecting on the matter a bit, Lucille decided that such an interesting-looking creature deserved a more dignified fate, perhaps on the mantelpiece. So, John climbed the fence and retrieved the carcass and they took it to town to get it mounted. As John tells it, the taxidermist took one look at the handful of punctured flesh and broken bones and said, "My God, it's a ferret. I can't touch it. They're an endangered species."

For years, a search had been underway throughout the West to see if any of these reclusive, nocturnal marauders of prairie-dog towns still existed. The longer the search lasted, the more it was feared they had slipped quietly into the abyss of extinction. And then John Hogg brought one to Meeteetse to get it stuffed.

73

It seems there are always antelope nearby when you're on the Pitchfork Ranch. These were close enough to photograph with a 70mm lens.

The taxidermist called the Wyoming Game and Fish Department. They traced the ferret to the prairie-dog colonies on the 120,000-acre Pitchfork Ranch, where the Greybull River wanders out of Yellowstone's spectacular Absaroka Mountains. As Jack tells it, the next thing he knew, the world was camped on his doorstep.

Green as a spring pea, backed by snow-capped peaks, with the cottonwood-lined Greybull meandering through its heart, the Pitchfork makes a pretty impressive doorstep to camp on.

The ranch was pioneered in 1878 by Otto Franc, a German immigrant who left a banana-importing business in New York to come west to improve his health and get into the cattle business. Butch Cassidy stole his first horse from Otto Franc.

Franc trailed his herd of Herefords into the valley of the Greybull even before the buffalo hunters arrived. Once there, he protected his interests aggressively. For that reason, he wasn't very popular with the sheepherders and home-steaders who followed. History buffs still hint that in that traditional animosity, so common in the settling of the West, lies the real reason behind the accidental shotgun wound Franc allegedly suffered while crossing one of his own fences.

After Otto Franc's demise in 1903, the ranch was purchased by a local family, the Phelps, which at one point included, by marriage, the well-known western photographer Charles Belden. While Belden lived on the ranch, photographs of the Pitchfork regularly graced newspapers around the country and appeared in *National Geographic* magazine. Promoted aggressively by Belden, images of the mountains and meadows of the Pitchfork became familiar to the nation and the world during the 1930s and 1940s. That notoriety spurred a thriving dude-ranch business that lasted until 1945.

Wildlife has been an important part of the picture along the Greybull for as long as we can trace the area's history. Evidence close by hints that mammoth hunters most likely roamed this area and, when Otto Franc arrived, the Shoshone and Mountain Crow tribes still hunted animals from buffalo to plains grizzlies in the valley of the Greybull. Today, herds of antelope browse the Pitchfork's meadows, and elk from Yellowstone use the ranch as winter range. Hunting has been closed down on the ranch a number of times— the first by Otto Franc during the late nineteenth century, after market hunters had wiped out the buffalo and decimated the antelope herds. Jack Turnell describes this episode as the beginning of the first game-protection association in Wyoming, which he says later grew into the state's Game and Fish Department.

In the early years of the twentieth century, the Phelps family had an active antelope-restoration program on the ranch. That program involved capturing fawns and bottle feeding them until they could fend for themselves. In this way, the precious few fawns that were produced each year were kept safe from predators, disease, and accident, so they could help rebuild the herd until it was once again self-sustaining. In an odd historical footnote, antelope fawns from the Pitchfork Ranch, bound for zoos in Germany, were part of the cargo of the zeppelin *Hindenburg* on one of its last voyages.

When the ferret that John Hogg had found was traced to the Pitchfork, it became apparent that the ranch was going to be involved in another wildlife recovery program, whether it wanted to be or not. Turnell remembers, "I knew I was under the gun. Media came from everywhere: San Francisco, Britain, Germany, Japan, ABC, PBS. People in droves wanted to come and look for 'em."

The former high school agriculture teacher was thus torn from a life of literally watching the grass grow to being forced to navigate the stormy seas of high-stress, crisis-oriented environmental politics. "Some people wanted us to kill all the bobcats and coyotes and raptors to protect the ferret," Turnell recalls. "Some said the cattle should be removed and tried to force me to admit I was doing something wrong. Others said this was the only place that there were any ferrets; maybe he's been doing something right."

"I said, 'Wait a minute.'"

Jack Turnell is a soft-spoken man, but he's firm in standing his ground when he knows he's right. To keep the orgy of attention from turning the ranch into a zoo and his family's life into a nightmare, Turnell closed the Pitchfork to everyone who wasn't directly involved in the recovery effort and joined with the Wyoming Game and Fish Department and the U.S. Fish and Wildlife Service to create an orderly process for satisfying

requests for access, studies, photographs, and press releases. Press releases had to be cleared by the advisory team, and an oversight panel was set up to review proposals for studies. Anyone who broke the rules "a couple or three times" found themselves "moved aside."

For a rancher, having an endangered species found on your place is a disaster of major magnitude—something akin to having all your cows come down with hoof-and-mouth disease, your grass polluted with nuclear waste, and your water all piped to Los Angeles. That's why ranchers whisper, only partially in jest, that the best tactic, if one does find an endangered species on their place, is to "shoot, shovel, and shut up." The rediscovery of the black-footed ferret could very easily have become a political shoot-out. It's a credit to Jack Turnell that it didn't.

Turnell handled what some would have considered a reason for total panic in what has become typical fashion for him. After helping to calm things down a bit, he started learning all he could about black-footed ferrets, even helping produce a movie, *The Mysterious Black-footed Ferret*, which was sponsored by the Audubon Society and aired by Ted Turner on his network, TBS-TNT. On the ground, Turnell proceeded conservatively. Sounding a bit like Yogi Berra, he concluded, "Obviously, whatever we'd done in the past, we'd done right, because we were the only ones with ferrets, so we decided early on that we'd just keep doing what we'd been doing all along."

After the ferret population reached a high (for the monitoring period) of 120 animals a couple of years after their discovery, the colony was suddenly infected with plague and canine distemper. Numbers plunged to a dangerous low of eighteen in the fall of 1985. With everyone holding their collective breath, a decision was made by the advisory team, which included Turnell, to capture the remaining animals before they completely died out. Hopes were that they could be treated and bred in captivity and that their numbers would increase to the point where some could be re-released. The program was a success. Vaccination bolstered the captive animals' immunity, and the population grew to 500 ferrets. Releases back into the wild are underway now, at locations away from the Pitchfork, where the prairie dogs are still infected. "If we get 20 percent survival on a reintroduction, we call it a success,"

"…it became apparent that the ranch was going to be involved in another wildlife recovery program, whether it wanted to be or not."

Turnell said. "We've got 20 percent at the first reintroduction site the first year."

Jack Turnell's date with environmental destiny changed his life in a way that he had never dreamed of, and if he had dreamed it, he probably would have blamed it on food poisoning or some other delirium-producing substance. Overnight, he had been transformed from a small-town rancher into an international environmental figure. He started getting calls to make personal appearances. He was invited to serve on boards. He started accepting the offers.

In addition to making him something of a celebrity, this abrupt change had a deeper effect on Turnell; it changed his outlook on the things that mattered most to him (after his family): life, people, and range management. "The ferret forced me to cooperate with people who I'd traditionally been an adversary of," he says simply. Turnell was referring to environmentalists, government agents, the urban press, all of whom most rural westerners look upon with suspicion, if not with contempt. "I realized then I could work with them and not feel threatened."

Jack didn't have to wait long to test this newfound confidence. While he was working with the ferret recovery team, he knew the next big environmental issue he would have to face was riparian zone restoration. Riparian or streamside zones are vital to wildlife in the arid West for a variety of reasons: living space, breeding areas, food, moisture, cover for hiding, shade in summer, and insulation in winter. They're also the areas most heavily used by livestock for many of the same reasons. Cattle tend to park in riparian zones, eating, defecating, trampling, unless someone moves them out. Ranchers and their livestock have been blamed for obliterating a significant amount of the West's riparian zones.

"I could see that train coming down the track," Jack remarks, "and I didn't want to get run over by it." He speaks softly as he loads up a slide show he has been putting together over the years on the ecology of riparian zones and the way they're managed on the Pitchfork. Jack has presented these slides and the talk he gives with them to dozens of groups around the West from the Greater Yellowstone Coalition to the Society for Range Management.

In 1982, he put the show together with the help of Quentin Skinner, a professor of range management at the University of Wyoming. That

"The ferret forced me to cooperate with people who I'd traditionally been an adversary of."

year the "train" Jack knew was thundering down the track hit with full force, and for a moment it looked as though it had run over him after all. The U.S. Forest Service moved to cut the Pitchfork's permitted cattle numbers on its 45,000-acre national forest grazing allotment a devastating 80 percent, from 1,250 pairs of cows and calves for ninety days to 500 pairs for forty-five days. The cut was mostly in order to try to restore riparian areas the agency said had been damaged by the Pitchfork's cows.

Jack knew he needed a crash course in riparian zone ecology to deal with this crisis, so he went to the University of Wyoming to see if the department's water wizard could help him out. He was wandering around the hallways of the Range Science Department when a man came up to him to ask if he needed help. Jack said, "I'm looking for Quentin Skinner."

"You've found him," was the reply.

Skinner is a professor of range management whose specialty is watersheds and wetlands. He listened to Jack's story and agreed to take a look and see what light he could bring to the situation. Skinner went out to the ranch, rode around to take a look, and talked to both sides. What he ended up telling the forest service and Jack Turnell was that they were both wrong. "Both sides were pretty ignorant of what that mountain was doing," remembers Skinner. "It's made of conglomerate. It's rotten. It's falling apart."

What that meant was that the cookie-cutter management approaches that both sides had brought to the mountain—the forest service and its ideas about what riparian areas should look like, and Turnell and his 100-year-old cattle management practices—left both parties standing on feet of clay, or in this case, the rotten remains of an old volcano. Skinner pointed out that the riparian areas on the Pitchfork were made up of volcanic conglomerate, a soft material that weathered and eroded rapidly and was being transported downstream at geologic superspeed. The image of stable riparian areas that the forest service had developed on granite mountains to the south did not fit the realities of the Absaroka Mountains/Greybull River drainage system. And while he was at it, Skinner pointed out that some of the impacts that were being blamed on livestock had actually been caused by elk and moose.

"An awful lot of social pressure is being exerted to achieve results that are in conflict with good science and the true state of our wildlands systems," states Skinner. "What society thinks is right in some cases could really be the wrong thing for the environment."

While Skinner was critiquing the forest service's position, he also pointed out that Turnell's management was based more on traditional grazing methods than considerations of what was good for riparian ecosystems. He suggested changes in management that would not only halt the degradation but would speed the healing. The technique most commonly recommended for restoring streambeds—fencing the cattle out—was not included in this list, and Turnell wouldn't have done it anyway. "That's just admitting bad management," he says. Instead, he began searching for a breed of cattle that had less of a tendency to loiter near water. He has found them, he says, in Salers (suh-lairz), a breed developed in the French Alps. He also began herding his animals out of the streambeds and adopted a rotation pattern for his pastures that rested each one every third or fourth year. Under Skinner's direction and Turnell's management, the wide gravelly bottoms of the Pitchfork's streams began to fill in with vegetation, the cutbanks began to round off and revegetate, and the stream channels began to narrow and deepen, especially in areas of slower flow and less cutting.

Another change Turnell made in his riparian management was to stop the traditional practice of trapping beaver out of the streams and dynamiting their dams. "Everybody used to say you had to keep channels open, so as not to restrict the flow of water to irrigation," Turnell explains. "Now we realize that if we slow the water down, the land holds more and the streams run longer. The Greybull River used to be dry part of the year past the ranch, now it flows year round." Today, beaver can be found just about anywhere on the ranch where there's running water, except in the irrigated fields. "Those fields are mine," Turnell states. "Whenever they get in there, they're in trouble."

Another tactic Turnell adopted for slowing down the water was flood irrigation. Turnell restored this historic form of irrigation to replace some of the sprinkler systems that had been brought into play in the hay meadows Otto Franc had cleared of sagebrush in 1878. The result was to turn the lowland meadows into huge sponges—domestic wetlands that captured spring runoff and slowly re-released it into the Greybull.

Taking all this into consideration, the forest service reduced the proposed cut in Turnell's allotment to 1,037 pairs for forty-five days. It was still a 40-percent cut. ("That's a lot of grass," grumbles Turnell.) But it proved to be a blessing in disguise because it inspired Turnell to change his management practices. He is now raising 542,000 more pounds of beef per year on his ranch, including 700 more cattle, in spite of the allotment reduction on public land.

There were blessings for the environment, too, and they're not disguised at all. They include greener, more lush, more prolific riparian habitats, more vital rangelands, less erosion, even scenery that is more natural. As part of his change in management, Jack did away with driving his cattle into the mountains to his forest service allotment in one large herd, as had been done since the nineteenth century. Instead, his cowboys drift them there, loosely, slowly, in groups of around 300. The result has been a decrease in the amount of trailing that scars the mountain slopes.

With so many management changes paying off, Turnell realized that the most basic reason for these successes was not new breeds of cattle or rest rotation systems. It was collaboration with people like Quentin Skinner and Phylis Roseberry of the Shoshone National Forest that changed these situations from problems into opportunities. Turnell realized that taking these issues out of either/or terms, and posing them in terms of shared goals (leavened with some good down-to-earth science), had turned what everyone thought was going to be a long fight leading to a shaky compromise into an opportunity for all sides to achieve what they wanted. As a result, the streambeds, wildlife, and rangelands on both sides of the fence, public and private, became healthier, and the ranch was more prosperous for it.

Believing he had stumbled onto something that could change the future of the environmental debate in this country, Turnell took his message of collaboration to a broader constituency.

First, he helped form the Wyoming Riparian Association and served as its president for a number of years. In this capacity, he began guiding tours of the ranch (with Quentin Skinner participating) for groups that included members of a diverse array of environmental groups and government agencies: the Sierra Club, the Audubon Society, the Wyoming Nature Conservancy, the Wyoming Outdoor Council, the Greater Yellowstone

"Believing that he had stumbled onto something that could change the future of the environmental debate in this country, Turnell took his message of collaboration to a broader constituency."

Coalition, the Bureau of Land Management, the SCS, the U.S. Fish and Wildlife Service—the list is pretty much encyclopedic. Turnell didn't just stop at Wyoming; he took his road show to Utah, Colorado, Montana, Missouri, and Washington, D.C. Ranchers' groups—the Society for Range Management, the Wyoming Stockgrowers, the National Cattlemen's Association—have also seen the slide show and talk. Some of them weren't quite sure what to make of it. "When I started talking about cooperation with enviros, a lot of my friends thought I was losing my marbles," Turnell jokes.

Turnell's faith in collaboration was so strong that, on days when he couldn't make it to a

Pitchfork Ranch cattle walk through stands of grass taller than their backs in wild meadows along the Greybull River. Much of this area was bare of any vegetation until rancher Jack Turnell changed his management of riparian areas under the direction of Quentin Skinner.

Riparian Association meeting, he let the Sierra Club representative vote his position. "She was familiar enough with what I stood for," he says, "and I'm confident she'd have let me do the same for her."

"Starting in the mid-eighties, Jack never let an opportunity pass to work out issues between environmentalists and livestock interests," claims Bob Oakleaf, nongame coordinator for the Wyoming Game and Fish Department in Lander. "I was gone 131 days on outreach last year," Turnell said.

Oakleaf believes Turnell manages to be effective in these circumstances in spite of the fact that he hasn't become a born-again environmentalist. "He's not the type who takes whatever environmental preaching is laid on him," states Oakleaf. "He can be abrasive." At the same time, while working with Turnell on the Black-footed Ferret Advisory Team (Jack is the only original

member still with the team), Oakleaf was impressed with Turnell's effort to become more than just a vested interest. "He devoted much time and effort to learn and understand ferret needs," says Oakleaf. "At the meetings it was very obvious that he was not there just to represent livestock interests."

In the field, Turnell's blunt approach can take some of his audience by surprise. In describing a typical dialogue on one of his riparian tours, Turnell tells me, "I get them on the creek and say, 'Here's what I want. Here's what I need to survive. What do you want?' Half the time they can't even tell me. It's not that they're not sure what they want. They just thought it would be a good idea to have a beautiful creek. So, I say, 'Tell me what you want, and I'll try to make it happen. And when I get it done, don't hammer me. Don't say cows are bad, because we fixed the problem.'"

In spite of his bluntness, or more likely because of it, one thing that stands out about Jack Turnell and his relationship with the people he meets in his outreach efforts is how many of those people come to genuinely like him. After I got back home from my visit to the Pitchfork, I received a package from Turnell that included a thick stack of letters of recommendation written in support of the Pitchfork's nomination for a number of stewardship awards. "I can't think of an award we haven't got," Turnell had told me. Most of these were letters from representatives of the same groups Jack has taken on tours on his ranch: in other words, just about all the groups there are in western Wyoming. And there were recommendations from others—governors, senators, representatives, government agencies, and educators both inside and outside Wyoming. A lot of the letters made warm references to Turnell and his tours. All talked about being impressed by what they saw.

Meeteetse, where John Hogg took the ferret to get it stuffed, is the town nearest to the Pitchfork. Actually, the ranch pre-dates Meeteetse by eighteen years. As one would expect from a large commercial enterprise near a small, isolated town, the ranch has always played a big part in the economic viability of the community. Recently, when the town's only general store, the Meeteetse Mercantile, was in danger of closing down, the Pitchfork bought a majority of stock in the store and Jack Turnell hit the bricks selling shares to community members to keep the store open.

The Greybull River and the Absaroka Range form a spectacular backdrop for the Pitchfork Ranch. In the foreground, cattle-holding pens that receive heavy impact at fall shipping time are full of rich, dark green grass after less than a year's rest.

As an additional way of supporting the community, the ranch has opened a museum of photographer Charles Belden's work to give tourists a reason to stop and spend some time in Meeteetse.

Lili worked with members of the local 4-H and Lions clubs, putting in more than 5,000 hours to build a boardwalk for downtown Meeteetse and give the town a general "sprucing up." Trees were planted and storefronts were cleaned and painted in an effort that left the town fairly glowing with the evidence of pride and community involvement. Meeteetse is still small and there are still some deserted storefronts in it, but even these are well kept.

A former teacher who has children of his own, Turnell also delights in doing things for kids. He recently built a lake on the ranch as an environmental study area for the local school system and a place for kids to go fishing.

While my wife, Trish, and I were visiting the ranch, we shopped at the Meeteetse Mercantile and saw old movies of the ranch and the antelope recovery effort at the Belden Museum. Out on the ranch, we took a photograph of Jack at his new lake and went along on one of his riparian tours. Turnell has given so many of these short tours in his car that he's starting to wear tire tracks in the grass on the off-road portions of the route.

The first stop on the tour is at a boundary fence between two pastures, where a small stream flows under the fence. When we were there in early August the left side of the fence was lush with willows and tall green grass. The right side had shorter grass and bare dirt rather than willows along the stream bank. As we approached, Jack said he has a standard question he always asks tour guests at this point. "Which side of the fence has had the highest concentration of cattle on it over the last few years?"

Before I could answer, Trish spoke up, "I know the answer to that," she said. "It's the side with the most stuff growing on it. The side with the willows."

"You know," said Turnell, "you're the first one to get it right."

"We may be very busy. We may be very efficient. But we will also be truly effective only when we begin with the end in mind."

STEPHEN COVEY
from The Seven Habits of Highly Effective People

DESERET LAND & LIVESTOCK, UTAH

Left to right: Bill Tobin, Mac Hedges, Rick Danvir, Bob Wharff, Jeff Gideon, Jay Olsen, Bill Hopkin, Lisa Richins, Mike Downing.

Not pictured: Greg Simmonds, Gary Jacobsen.

CHAPTER TEN

DESERET
LAND & LIVESTOCK

Bill Hopkin, Rick Danvir, Greg Simmonds

UTAH

*On Wildlife and Lots of It, Sage Grouse Eggs, Riparian
Areas, and Moving Elk with Bird-watchers*

O N A RAINY MORNING IN LATE JULY, I was driving north out of
Evanston, Wyoming, hoping to outrun a stormcloud that threatened
to make the day difficult for photographs and field trips. Gliding past
the rolling seas of sagebrush that cover the valley of the Bear River,
I crossed the border from Wyoming into Utah and turned down the
road marked "Deseret Land and Livestock."

Along the broad valley of Saleratus Creek, I had to work to keep my van from
swapping ends on road clay turned to slime by three days of rain. As I fishtailed my way
toward the ranch headquarters, I was greeted by some of the main reasons I had come
to the Deseret: In the tules, near the creek, a flock of sandhill cranes foraged; overhead,
a marsh hawk cruised; and, in the distance, a golden eagle flew a search pattern over the
sage flats. Standing in the grass at the foot of a string of low hills, a small herd of antelope
raised their heads and pricked their ears as I passed, slipping and sliding by. The Deseret
is known throughout the West for the innovative ways in which it has been able to combine cattle ranching and wildlife management in a comprehensive program of stewardship.
When you pay a visit to the ranch, the fruits of that work are not hard to see.

The Deseret Ranch was founded in 1870 by a group of Mormon sheep men. The
name they chose for their ranch, "Deseret," comes from the Book of Mormon, in
which it is used to refer to the honeybees that a persecuted people took with them into
the wilderness. For Mormons (members of the Church of Jesus Christ of Latter-day
Saints), "deseret" has come to be synonymous with industriousness and hard work, and

the name is used extensively by the Church and its members as an expression of the high value they place on these virtues. It even serves as a basis for the nickname of the state of Utah, "the Beehive State."

"I don't think there are good and bad animals, just good and bad managers."

The Deseret Ranch has changed hands a number of times since it first came into existence. In the late 1970s and early 1980s, it was owned by a wealthy Hong Kong entrepreneur of Dutch trader stock, Joseph Hotung. Having spent much of his life in one of the most densely populated places on earth, it is easy to imagine Hotung buying this collection of almost endless sagebrush flats, broad valleys, and low mountains—at least partly for the sheer fascination of owning a piece of land so big, so uncrowded, so free, and so private. All but a few of the Deseret's 201,000 acres are owned by the ranch. Unlike a lot of rich absentees who get into ranching for the romance and the tax write-offs, Hotung was strictly business. When the Deseret began losing money, he told the people he had hired to manage it, "From now on, you're on your own. If you lose money, I won't make it up. If you make money, it's yours." That mandate sent some folks packing and inspired others to take the risk and run with it. Full speed ahead, they started putting together a plan to operate the ranch the way they thought it should be run—holistically,

sustainably. When the Mormon Church bought the ranch in 1983, it kept the management team intact, though it included people who weren't Mormons. The Church knew a good thing when it saw one.

The spark plug of that management team was Greg Simmonds. Simmonds was a graduate student at Utah State University when the man who preceded him as the Deseret's manager came seeking his services as a consultant. "He was thinking about putting in seventeen more irrigation sprinklers," relates Simmonds. "And I was working on a thesis on how farming (irrigation and crop raising) affected the efficiency of Intermountain cattle ranches—it killed it."

Simmonds showed the ranch that they not only didn't need the increase in technology but that they could do better with less. The solution was as surprising as it was mundane. Simmonds increased the ranch's overall efficiency by improving its system of accounting and accountability. The techniques he used are the same that many of us are using to bring order and effectiveness to our lives: find out where you are, decide where you want to go, and plan how to get there. Simmonds's techniques are a mixture of business guru Steven Covey's *The Seven Habits of Highly Effective People*, Alan Savory's *Holistic Resource Management*, and his own.

"Covey taught me that the human factor is where the highest potential lies," explains Simmonds. "That humans can be ultimately destructive or extremely creative, and when the latter is the case, that one plus one equals more than two." This new math of abundance, according to Simmonds, is the antithesis of the old math of scarcity, which "tells us that the only way I can have more is for you to have less; that one plus one always equals less than two."

For most ranchers, the math of scarcity is as real as a family that wants to eat and enjoy the standard amenities of modern life, or a loan payment that needs to be paid on time. Since livestock represents livelihood to them, a mouthful of grass that goes down an elk's throat looks for all the world like a portion of food that won't find its way into the mouths of their family. In a world seen through those eyes, native animals, such as deer, elk, antelope, even prairie dogs, rabbits, and mice, are in direct competition with wives and children for food and sustenance.

Simmonds sees wildlife as the opposite of a threat. He sees it as an essential part of the complex web of beings and interactions that makes up the vital organs and life functions of the ecosystem upon which the ranch, its people, its livestock, and even the surrounding community depend for survival. Although some of the other people on the ranch don't see wildlife as quite so essential, at least they see it as unavoidable. "If your land is in good shape, you're going to have lots of wildlife, so you might as well figure out how to deal with it," one of them told me. To figure out how to deal with wildlife and have the Deseret function as near its highest potential as possible, two full-time wildlife biologists and one half-timer are included in the ranch staff. Their work is supplemented occasionally by others: a graduate student from Belgium working toward her Ph.D. by studying the foraging habits of sage grouse and a professor from Utah State University studying prairie dog populations. In addition, the ranch serves as a site for studies by other institutions and groups attracted to it by the innovative and open atmosphere.

Rick Danvir became chief of the ranch's wildlife division in 1990, after having started out as a biologist under the previous wildlife manager. His position is one that isn't even dreamed of at most other ranches. Though the hair that peeks from under his hat is graying, Rick is gangly and slouched with the loose walk and relaxed mannerisms of a lifelong teenager. When I arrived at the ranch, the first thing he said to me was a good-natured complaint, in very explicit language, about a sewer repair job he had ended up having to do. It was exactly what the doctor ordered when it came to letting me know the proper way to relate to a ranch that was owned by a church—no sweat.

Rick has worked as a wildlife biologist in three states: Utah, New York, and Colorado. "I came out of the typical wildlife managers mold," he confesses. "I blamed livestock for a lot of the problems with wildlife." Now that he's on the other side of the fence, Rick sounds like someone who has had his horizons expanded, not someone who has sold out. "It was an eye opener to me," he admits. "Now I don't think there are good and bad animals, just good and bad managers."

The studies the ranch has done to make sure it comes down on the side of good managers are strictly nuts and bolts and visionary at the same time. To test the truth of claims that intensive grazing can turn the eggs of ground-nesting birds into a cow-stomped omelet, ranch biologists placed a number of chicken eggs in simulated sage grouse nests in an area where cattle were concentrated more densely than they ever are in normal range situations. "We had a cow and a calf to

Some of the 2,000 elk living on the Deseret Ranch put a little distance between us and them on a day when the temperature dipped to forty below. The Deseret Ranch is an excellent place for wildlife watching and photography any time of the year.

every two acres in there," explains Danvir. "There were cow tracks on every square foot of ground." In spite of that, only one egg was stepped on. "We found out that, unless they're stressed, cows walk around the sagebrush under which sage grouse lay their eggs."

The Deseret Ranch can afford to spend as much time and effort as they do on wildlife because they don't just talk about them as an asset, they actually treat them as one. "Wildlife pays for itself on the ranch," Rick states. By that he means wild animals that live on the ranch produce sufficient income to pay for the salaries of the biologists who study them, the time and materials needed to fix the fences they bust, even the forage they consume. And when Danvir says that, he's not just guessing; he knows, or at least he has an idea based on more than a seat-of-the-pants estimate.

Ken Clegg is a student at the Utah State University whose research is funded in part by the ranch. He spends many of his days looking through a powerful spotting scope watching elk eat grass. As they graze, he counts the number of bites they take and uses that number to come up with an elk-grazing rate in bites per hour. Then he goes to the area where the animal he has been watching has been foraging and, by comparing grazed and ungrazed plants, gets an idea of how much grass an elk consumes when it takes a bite. By extrapolating this over a full year and multiplying it by the number of animals on the ranch, Danvir can calculate what portion of the forage that grows on the Deseret is consumed by elk. The same works for other animals.

This bit of knowledge comes in handy in a couple of ways. It helps the ranch calculate how much money must be produced by hunting and nonconsumptive programs, such as wildlife photography and bird-watching, for wildlife to not be a liability. It also helps the ranch plan the most efficient way to use the very limited amount of forage that grows there in just fifty frost-free days a year, on an average of seven to eleven inches of annual moisture.

Conventional range science and ecology tell us that wildlife and livestock are in competition for available feed. In order for there to be enough forage for wildlife, scientists, bureaucrats, and activists tell us that livestock must only consume so much. The whole enchilada minus X for cows leaves Y for elk and other things. This too, says

Simmonds, is an example of the old math of scarcity. The management of the Deseret has found that, by using information about the ranch's forage production and about the way the various animals use that forage, they can let cows and elk belly up to the same table in the right way and at the right time and end up sharing an enchilada and beans, maybe with a side of chips. There are examples of this.

Both elk and cattle prefer their grass green and lush, the way you find it along stream banks. For that reason, both like to spend as much time as possible in riparian areas, and both can have a devastating effect on them. The situation doesn't sound natural because it isn't. Before there were ranches, predator control, and hunting seasons, a herd of elk that stayed in one place soon attracted unwanted admirers—if not wolves, a mountain lion, or a grizzly bear, then human hunters eager to take advantage of the easy pickings. If the elk didn't move they were killed. With most of the natural predators killed off and humans hunting for sport a few weeks out of the year instead of pursuing the animals year-round for subsistence, elk are more prone to park in one place and eat until they cause real damage.

Since the Deseret has taken measures to keep its cattle from overgrazing riparian areas, ranch personnel feel just as moved to keep elk from doing the same. Since cattle and elk generally don't graze together, especially when there are lots of cattle, Danvir decided to use cows to move the elk.

"When we put the cows in here, the elk were all down in the canyon bottom, the grass was beat, and they were really hitting the browse," Rick recalls, as he and I look down a long narrow valley with a ribbon of willows at the bottom of the V and grass giving way to sagebrush higher up on the sides. "The cattle headed right for the riparian area," he continues, "and they pushed the elk out and up the mountainsides. When the cattle found the bottoms already heavily grazed, they followed the elk up the slopes, forcing the more mobile wild animals higher and higher until they had been completely displaced. Two weeks later, after the cows had grazed the forage off the slopes they, too, were moved out."

So far, so good. The surprise came four weeks later, when Rick went back to check on how the area was recovering and found three times as many elk as had been there when he

"With most of the natural predators killed off...elk are more prone to park in one place and eat until they cause real damage."

decided to move them in the first place. This time, however, the animals were dispersed over the canyon slopes, eating the regrowth that had sprung up where the cattle had mowed down the dry material left from last season's growth. Not only had the cattle moved the elk out of the riparian area, but they had actually increased the amount of feed available to the more finicky elk.

"We found we could create an elk magnet with cows," Rick tells me. But what do you do with an elk magnet when the problem is too many elk? You use it to disperse their impact over a larger area to keep them from overgrazing the riparian areas. In the high desert, where healthy riparian areas act as storage reservoirs for scarce moisture, that is an extremely useful tool.

Rick and his staff are now investigating other means of moving elk, using modern recreationists to do the job once done by Stone Age hunters. "We've thought about opening problem areas to fishing or wildlife watching," says Rick. "We hope that if we bring in the people, they'll move the animals."

Most of the money to support this sort of study and innovation comes from the Deseret's extensive hunting program, although the ranch is working to develop a number of nonconsumptive programs as well. In this latter effort, the ranch has solicited interest from professional wildlife photographers and writers by providing complimentary visits to blinds near sage grouse strutting grounds. A huge old irrigation reservoir built in 1905, which attracts waterfowl, and a number of riparian areas, some complete with beaver dams, also provide good bird-watching. And the Deseret offers unmatched opportunities for viewing larger animals, such as elk, mule deer, and moose. When I was there, I saw more animals and a greater diversity of species in two days than I did in two weeks at Grand Teton, Yellowstone, and Glacier national parks combined.

Because of the Deseret's exemplary wildlife management program, the State of Utah Division of Wildlife has made it a posted hunting unit. That means the ranch is authorized to sell permits to hunt on its land and to limit hunters to those who hold the permits. As part of the deal, the ranch returns to the state 10 percent of the funds it receives for male deer, elk, and antelope permits, and all of what it receives for antlerless hunt permits. In addition, the state is allocated 20 percent of the premium elk permits, which are in

highest demand and command the most money. To give an idea of how much demand there is for hunts on the Deseret, each of twenty premium guided elk hunts the ranch sold in 1993 went for $8,000. More than 1,100 people entered the lottery for the five state permits.

As successful as this hunting program is, the ranch views it as more than just a way to control wildlife numbers and produce money. It is part of a well-thought-out management plan that must achieve predetermined goals with a minimum of negative impacts, the same as any other management program on the ranch. For instance, the hunting program is designed to keep the animals that make up the herds living on the ranch as diverse and healthy as possible. "Big bull elk and big buck deer are the ones everyone wants to shoot, so we found hunters were leaving less healthy animals as breeding stock," Danvirs relates. To help reverse that trend in negative selection, Rick came up with what he calls a management hunt. "We trained the guides to pick animals that are still good but are never going to be the best, and we charge less to hunt them. Now it's our most popular hunt." To make management hunts even more attractive, and to use them as a way for hunters to become better acquainted with the ranch and its methods, Rick also made participating in them one way to accumulate points to qualify for the premium permits that otherwise are issued by lottery.

To control herd size and keep the gender mix from becoming unbalanced, the ranch also has a doe deer and cow elk hunt. "When we need to remove females, hunters who are on the ranch are required to participate," Rick explains.

All this sounds terribly premeditated, manipulative, and cruel in this day when we are questioning whether or not Homo sapiens should hunt at all. But our forebears have been interacting with animals in this way for millions of years. If we humans cease to fulfill our traditional ecosystem role as predators, can that be called anything less than a significant change in the natural workings of things? If the ecosystem can suffer from a lack of wolves, can't it also suffer from the lack of the animal that has been the primary predator for millennia—humans? If functioning as a part of nature rather than serving as lord over it is essential to environmental balance (as many of us say it is), the Deseret seems a lot closer to that goal than places where there is no hunting at all and

"Since cattle and elk generally don't graze together, especially when there are lots of cattle, Danvir decided to use cows to move the elk."

Rick Danvir and Ken Clegg check a thick stand of grass and sedges in a drainage that once was badly eroded. New growth stimulated by alternating impact and rest in the bottoms of these gullies is now filtering so much sediment out of floodwaters that the arroyo is being filled up instead of cut deeper.

humans relate to animals in a way that the eco-system has never before experienced. One thing is certain: it's hard to argue with the Deseret's results. Within the last twenty years or so, the elk herd that lives on the ranch has increased from 350 animals to more than 2,000. It has become so large, in fact, that it has begun to spill over onto neighboring ranches and to sour some of the Deseret's community relations.

"Our wildlife program isn't popular with some of the folks in the local community," says Bill Hopkin, who took over the job of ranch manager when Greg Simmonds was promoted to manage a bigger piece of the Church's growing stable of ranches. "A neighbor could go to bed at night with a nice big haystack outside and wake up in the morning with nothing but a bunch of well-fed elk cleaning up the leftovers." Hopkin

is a fire plug of a man, with prematurely gray hair that belies his rugged youthfulness. He knows the concerns of his neighbors. He grew up here and is related to many of them.

To keep the elk from wreaking havoc on the community and acting as ambassadors of ill will, the Deseret has begun feeding the animals over the two hardest months of winter. Usually, about 1,000 elk show up. In the winter of 1993—an exceptionally hard one—1,500 animals consumed more than 900 tons of hay, most of which was purchased alfalfa, which the elk prefer over the home-grown grass hay the ranch feeds to its cows.

Hopkin and Danvir are both quick to point out that the ranch's mission—to make a profit while at least maintaining the quality of the ranch's environment—must be carried out while keeping good relations with the community.

Because of that, and because rural society and the Mormon Church have always stressed community harmony, the ranch is very sensitive to the feelings of its neighbors. One study done on the ranch by Utah State University has confirmed that cattle and wildlife can graze together in ways that are mutually beneficial. "That'll help," says Hopkin.

Along with its success with wildlife, the Deseret is known for what it has achieved in rehabilitating riparian areas, especially on the part of the ranch that is sagebrush country. Here the land is low and dry and what little water there is comes as rain or snow and leaves in a muddy hurry. For years before the current management team took over the ranch, these washes were little more than gravel-bottomed storm sewers.

Under the Deseret's careful management, the bare gravel that once served as the floor of these washes has been covered by a dense turf of grasses and sedges. In places, that turf has grown so thick that the waving leaves of the plants lean over the stream and touch above the ribbons of clear water that flow beneath the shaded banks. The plants serve as fibers to filter sediment from the periodic flash floods. In one place, this has resulted in filling the arroyos to such an extent that it has nearly buried a set of four-foot fence posts that were driven into the streambed before the management change.

While I was visiting the ranch, I took field trips with Bill Hopkin, Rick Danvir, Ken Clegg, and another ranch biologist, Bob Wharff, to several of the areas that have changed dramatically under the Deseret's management. When we visited these riparian grasslands, they were glowing with vitality and spiced with wildflowers. In a few places, Rick pushed aside the grass to show that willows were beginning to sprout. "This has taken a long time," he said. In other places, we could see rotting skeletons of sagebrush. These arid lands plants had been drowned by the rising water table of moisture absorbed and held by the natural sponge of sod, roots, and soil created by riparian regeneration.

The Deseret has managed to restore these desert riparian areas while continuing to use them for grazing. As you might suspect on a ranch where holism is a byword, that's not an accident. Where cattle and sheep once were a liability, they now serve as the agents of restoration.

One evening, as the mountain shadows were beginning to stretch across the foothills, Rick

Danvir and his father took my wife, Trish, and I on a drive up into the mountains to see how much wildlife we could see. We were more successful than I expected, enough so that my wife, who is an environmental educator, began to think that maybe I had not gone crazy to start writing about ranchers as environmentalists. We saw moose, beaver, mule deer, elk, coyote, nighthawks, sage grouse, poorwills, an unidentified owl, and several other raptors.

As we sat at the top of a steeply descending valley with a dozen or so beaver ponds stairstepping like fertile terraces down the valley floor, Rick summed up what he believed is the Deseret's greatest achievement.

"We've made the land a little better, and I'm proud of that, but if we are a success story, one way we're a success is this: You take Jeff Gideon, the cow boss; he's a cowman all the way, but when we're planning something we're going to do with our cattle herd, he'll say, 'but what's that going to do to wildlife,' and sometimes our biologists will ask what effect a wildlife program is going to have on our cattle-grazing operation.

"I've been to plenty of meetings," Rick continued, "where the rancher hates the wildlife guy's guts and the wildlife guy hates the rancher's guts, and they'll both do anything they can to keep the other one from getting in their way. Their fight has nothing to do with the cattle or wildlife or with the condition of the land."

Then he thought a minute and said, "And at the same time, it has everything to do with it."

"If we humans cease to fulfill our traditional ecosystem role as predators, can that be called anything less than a significant change in the natural workings of things?"

*"We know that many fundamental beliefs of modern science arose
as heretical speculations advanced by nonprofessionals."*

STEPHEN JAY GOULD
from Ever Since Darwin

TIPTON RANCH, NEVADA

Front row (left to right): Tony and Jerrie Tipton.

Second row (left to right): Tommie Martin, Ron Martin, Tom Lamb, Tom Frolli, Ralph Young.

Third row (left to right): Jeff Weeks, Betsy MacFarlan, Norm Cardoza.

*Back row (left to right): Duane "Swede" Erickson, Earl McKinney, Rich Benson, Ted Angle,
Roger Johnson, Sydney Smith, Joe Maslach, Doug Busselman, Dayle Flanigan.*

Not pictured: Steve Rich, Barbara Curti, Larry Teske, Roger Mills.

CHAPTER ELEVEN

THE
TIPTON RANCH

Tony Tipton, Jerrie Tipton, and Tommie Martin

NEVADA

*On Old Dogs and New Tricks, Getting out of the Cow
Business, Diversity, and Having Babies*

I'M OUT OF THE CATTLE BUSINESS. I'm into the land management business,"
declared Nevada rancher Tony Tipton, as he leaned back in his chair and smiled
one of those smiles that tells everyone he's laying it on the line.

Tipton's comment was directed at about a half dozen federal land managers, a
couple of urban environmentalists, and a representative of the Nevada Cattleman's
Association crowded into the cramped little kitchen of the sheepherder's cabin Tony
and his wife, Jerrie, use as a home and ranch headquarters. None of them objected to
what they had just heard, although all of them knew that, at that very moment, 500 or
so cattle, about a dozen horses, and a little less than a hundred goats were grazing on
Tipton's 26,000-acre forest service grazing allotment at the foot of the Toiyabe Range
near the town of Austin in Central Nevada. No one objected because they knew the
Tiptons weren't trying to deceive them, and because they had witnessed enough sur-
prises at the hands of Tony and Jerrie Tipton to know that they didn't say things just
for the effect of saying them.

What made Tony Tipton's claim outrageous was not that he was the first to make
it, or even that he was the first to pull it off. Ranchers have been saying as much for
decades: that they are stewards of the land, agents of restoration, wildlife's best friend.
But those claims had been viewed as little more than self-serving proclamations, added
entries in the endless string of bumper-sticker slogans and one-liners into which this
issue has degenerated: "For a rancher, every day is Earth Day," "And on the eighth day,
God made ranchers." Peeking out from the mud spattered on a pickup bumper, such

The stream banks of lower Big Creek, once mostly bare dirt and gravel, are now covered with living diversity. The Tiptons achieved those results not by fencing livestock out of the area, but by changing the terms under which the animals were brought there.

animals, thus, have become a tool for improving the land rather than a commodity produced upon it—a tool unlike bulldozers and bureaucrats, with the advantage of being able to pay its own way by creating cash flow as it restores. It was this paradigm shift that Tony Tipton was referring to as he addressed the annual monitoring meeting of the TWWMT.

Such revolutions aren't won easily. They're carried through by skillful politics and hard work. The Tiptons know how to use both, but in this case they succeeded mostly by means of the latter. To do so, they first had to convince district ranger Dayle Flanigan, of Toiyabe National Forest, Austin, that livestock could be something other than a scourge upon the land, more than the shit-smeared, cloven-hoofed, foul-smelling locusts that writer Edward Abbey raged about.

It wasn't long ago that Tony Tipton seemed like the last man on earth to be promoting understanding between ranchers and the federal government. In the late 1970s, in the days leading up to the Reagan Revolution, Tipton was a radical, a trench fighter in the war against environmental regulation and government control. He was one of the most aggressive of those "Sagebrush Rebels," who sought to free the public lands of the West from the heavy hand of government agencies and urban interest groups, which they saw as no more than thinly veiled colonial powers. With the build of a welterweight and a mind and temper as quick as the hands of anybody named Sugar Ray, Tipton wasn't above slamming someone against a wall to make his case. "He was known as a real hothead up in Winnemucca," remembers one BLM staffer who had dealt with him there. "When we heard he was moving down here to Austin, nobody was looking forward to it."

By the time the feisty range rebel arrived in Austin, however, he had changed his stripes. Instead of bouncing adversaries off barns, he began inviting them to his ranch to ask how they

declarations come across as little more than clutching at straws to keep the ship from sinking.

What made Tony Tipton's declaration more than a bumper sticker was that he proved he could deliver on it and he got someone to believe that he had proved it. He even had a signed memorandum of understanding (MOU) with the U.S. Forest Service to prove it. That MOU stands as evidence that the Tiptons have taken their land-management success beyond slogans and bluster. They've even taken it a step beyond good stewardship; for a steward, after all, is a caretaker—a maintainer of the status quo. The Tiptons and their team have transformed animal management into a practice that has been recognized by the U.S. Forest Service as a legitimate means of making nonfunctioning ecosystems work again. The MOU has transformed the Tipton operation from cattle ranch to the Toiyabe Watersheds and Wetlands Management Team (TWWMT). The Tiptons' thought he should run it. The reason for the transformation was as much self-interest as change of heart. During his sagebrush rebel days, he had experienced directly how overwhelmingly more powerful urban interests were than rural peoples. To Tipton, that revelation made it clear. In spite of ranchers' nearly legendary reputation for winning political battles, unless they changed, they would inevitably lose. While some of his old compadres kept up the good fight, Tipton decided that the only way he could keep doing what he wanted to do—ranching and making a living off it—was for him not only to join his opponents but to get them to join him.

In 1987, Tony, Jerrie, and Jerrie's sister, Tommie Martin, a conflict resolution facilitator, sent out more than 200 invitations to representatives of Earth First!, the Nature Conservancy, Sierra Club, the Nevada Cattlemen's Association, and other interest groups, as well as to all the government land management agencies with an interest in the more than 40,000 acres of public lands that are grazed by the Tiptons' cattle. All were invited to come to the ranch to see if they could agree on a set of common goals and commit to work together to achieve them.

The response was sparse at that first meeting (no environmentalists and only a few government staffers showed up), but still it resulted in the formation of TWWMT. Seven years later, TWWMT is still meeting at least four times a year, bringing

ten to twenty volunteers together to review the team's goals, to discuss the plans they've formulated to achieve those goals, and to monitor the progress that the ranch has made toward them.

On June 4, 1993, I joined members of the team at the ranch headquarters south of Austin for one of their regular meetings. Outside, a late spring snow was falling and the 11,000-foot Toiyabe Range was cloaked in white. Where we were, at 6,000 feet on the Reese River Valley floor, the snow melted as it touched the ground, tinting the hay meadows surrounding the old cabin with the iridescent jade of reawakened growth. The chill moisture also awakened the Tiptons' flock of goats, inspiring them to clatter over the government pickups and private four-wheel-drive vehicles parked outside in a noisy game of king of the hill. In their game, the goats proved themselves to be true egalitarians. No vehicle, old or new, was denied its turn.

Inside the cabin, the group that had witnessed Tony Tipton's declaration was sheltered from the cold and damp. Drinking coffee, they offered greetings to new arrivals, who stomped and shook the way a wet dog shakes as they stepped inside the door. Some had driven from as far away as Reno, 340 round-trip miles to the north, taking a weekend and then some, to participate in three days of biological monitoring, ranch tours, and informal discussions.

The Tiptons organize these get-togethers according to one of the cardinal principles of ecology: that diversity creates strength. When invitations are sent out, no one is overlooked because their views are too radical. Opponents of grazing as well as its friends are invited. Among those who get a regular invitation is Rose Strickland, the Sierra Club's leading activist on grazing issues and a vocal critic of public-lands ranching. Strickland, however, has never attended, explaining that the Tiptons' ranch is a small part of the rangeland reform picture, and that Austin District Ranger Dayle Flanigan has the situation under control.

Dan Heinz also gets a regular invitation. He is a field representative for American Wildlands and an occasional consultant on grazing issues for the Sierra Club and National Wildlife Federation. Heinz has visited the Tiptons' ranch a number of times, but never while a team meeting was underway. "I've never seen consensus management work," he says.

"The Tiptons organize these get-togethers according to one of the cardinal principles of ecology: that diversity creates strength."

The Toiyabe Wetlands and Watershed Management Team maps a stream profile on upper Big Creek on Tony and Jerrie Tipton's forest service allotment.

Dayle Flanigan has a different opinion of the value of TWWMT. "If I get a letter of concern from someone about this allotment," he reports, "I just tell them when our next team meeting's going to be and say, 'Why don't you plan to attend?'"

Carrying the environmentalist banner on this day are Adrianne Kovacs and Craig Downer of the Wild Horse and Burro Association. Kovacs has attended a number of meetings; it is Downer's first. Most, if not all, of the others present, would identify themselves as environmentalists as well, including the agency staffers (as much as their positions as government employees will allow). They belong to the groups, read the newsletters, share the concerns. To some degree, so does the cattlemen's representative, Betsy Macfarlan. Macfarlan likes green grass and clear streams. She just likes them with cows on them—at least some of the time.

Meetings of TWWMT are normally facilitated by Tommie Martin. Tommie is a modern-day circuit rider who pilots her Chevy Suburban around the West, wearing out engines as fast as missionaries wear out welcomes. She is on the road more days out of the year than she dares to add up, spreading the gospel of teamwork and goal-oriented collaboration to anybody fed up enough with confrontation to listen. Because she has just finished facilitating a session at a ranch in Arizona and has been driving all night, she calls and says she has to take a break and will be a little late. The format of these meetings has become something of a ritual, so we decide to start without her.

Who are you? Why did you come? And what goal would you like to achieve by being here? Those are the three familiar questions each person is asked to answer in turn. Some take a few seconds; others take considerably longer. Most people have made these presentations many times before. They breeze through it, laughing, telling inside jokes. Other participants are obviously newcomers. They struggle. If someone says they don't have much to say, you know it's going to be a stem winder.

In most cases, the reason for being here is a variation on two themes: to take advantage of this opportunity to learn more about the land and how it responds to creative management and to express a commitment to solving problems by working with others rather than struggling against them.

"... how do you get cows to make grass grow when they're almost universally recognized as one of the planet's most effective agents at making it disappear...?"

Tony Tipton's turn comes last, and that's when he makes his claim about being a land manager instead of a rancher.

The Tiptons weren't always so confident. Just as Dayle Flanigan had to be convinced, so did they. When they got their team together, they knew they were going to have to prove that there were more reasons to keep coming to these meetings than smiles and a free lunch. They knew they were going to have to produce some sort of demonstration, a sign that what they were doing went beyond wishful thinking and touchy-feely words about how collaboration was more productive than conflict. And it was going to have to be dramatic.

To provide that demonstration, they picked the most challenging test they could find. It stood in the form of a steep, barren pile of clay, which formed a dam that held back a settling pond for a gold mine located on the same forest service lands that the Tiptons graze. The dam was a scar visible for miles, including from the Tiptons' headquarters. It was constructed in the early 1980s to hold the soup of chemical solvents used to concentrate the mine's gold ore to make it easier to truck off to a smelter. To construct the dam and the basin behind it, thousands of tons of a white clay named Nevada gray were gouged from the sagebrush-covered slope and shaped to form a bowl-like depression. The surface of this breastwork was then compacted by a bulldozer into armor-clad hardpan, and what was supposed to pass for topsoil was spread over it, as a miser might have spread caviar over a hillside. In the near decade that had passed, erosion had removed most of the topsoil and started to cut into the dam itself. All that had grown was a few tumbleweed, and even they were slowly disappearing from the scene.

Tony and Jerrie Tipton offered to cover that eroding scar with green living grass, using only cattle and hay and enough fossil fuel to haul the hay into place to do it. The stakes were high. If they succeeded, they hoped to convince the forest service and their management team that their claims that livestock could be used to improve the land were not just self-serving hot air. If they failed, their team members would most likely fade away and find something better to do.

But how do you get cows to make grass grow when they're almost universally recognized as one of the planet's most effective agents at making it disappear—and on such a hostile environment? The pile of Nevada gray had never supported anything but dust devils. To effect a change that sounded like magic and worked like common sense, hay would be spread over the surface of the dam, enticing the cattle to climb its steep slopes. There, they would eat and tramp, defecate and urinate—fertilizing, tilling, tamping, and planting the seeds of the hay and the five-way wheat grass mix that had been broadcast in case the hay seeds weren't enough.

The idea was simple, but the work was brutal. Thirty-two tons of waste hay had to be spread over nearly ten acres of land sloped so steeply that you almost had to crawl to get up it. The amount of hay was huge and the quality low, to ensure that the animals wouldn't eat all of it but would churn much of it into the dirt. Along with their dung, that debris would create a mulch to hold moisture and fertilize the soil, thereby creating the conditions for growth.

For four days in November 1989, these siblings of Sisyphus—Jerrie, Tony, and Tommie— with the help of a hired hand and a few team members, skidded 100-pound bails of hay down that slope, slipping, sliding, and falling as they scattered the organic material over the surface. Then they climbed and crawled back up to do it all over again, cussing cows and their own stubbornness as they went.

Their neighbors told them that it wouldn't work, that the old bulls and young calves that made up much of their herd wouldn't climb such a steep slope, and that the animals would break their legs if they fell. In typical Nevada fashion, the neighbors placed bets, sitting in their pickups, watching. For the Tiptons, it was damn the doubters; full speed ahead. From daylight till dark, they kept a steady stream of hay spreading across the dam site. The cows did their job: eating the fodder and stomping and grinding the waste into the Nevada gray. And when the work was finished, the slope was broken by innumerable terraces cut into the hillside by the passing of the animals, and the surface was thatched with trampled-in debris.

Then the cows were removed and the land was permitted to gestate. Everyone waited.

When spring came, all those with a vested interest—the Tiptons, the team, the bettors, almost everyone in Austin—watched the white scar on the mountain slope. Ever so slowly, but with surprising speed, it turned to brilliant green.

"It was so green, like Emerald City in *The Wizard of Oz,*" remembers Tommie. "And the locals started calling it that." Bets were paid and, by the end of summer, the Tiptons and their team waded through the thigh-high grass, clipping and weighing samples of it, as they would to estimate the production of a hayfield. Their tests revealed that their cow-cultivated mine site had produced more grass than some of their neighbors' irrigated hayfields—and had done it on less than six inches of moisture!

When I visited the mine site four growing seasons later, Jerrie Tipton urged, "Climb up to the top, I want you to see something."

I scrambled up the slope and noticed that the grass was already beginning to thin. For want of the power that made it, say the Tiptons. After the cows had made the grass grow, the area was deemed too fragile for them to return, and the forest service and the mining company went back to applying the only natural tool they have for maintaining the land: rest. Looking at the grasses on the slope, you could count the years since the Tiptons and their cows had been here by the steadily shortening stems of the grass plants. Blades of grass that had grown the first year had been mummified by the Great Basin winds, but they were still higher than my knee. This year's growth, which had just formed seed heads, was barely as tall as a ruler and clung to the edges of plants whose center was still occupied by the corpses of years past.

As I climbed, I stopped to catch my breath a couple of times. Gravity tugged at my legs. It really was steep. Once at the top, I turned and asked what it was that I was supposed to see.

"I wanted you to see what it felt like to walk up there," Jerrie answered. "I did it hundreds of times in those four days."

She was right. One can never fully appreciate work by merely looking at its results. But one trip up that slope only gave me the barest hint of what it must have been like to struggle to the edge of endurance to prove a point; not just any point, but one that was seen as vital to being able to continue a way of life. It must have felt like being in a war or running a race that you know you have to finish, no matter how tired your legs get. No matter how many excuses you can think of.

And there were plenty of excuses at the top of that pile of clay. Looking across the immensity of the Reese River Valley and the sagebrush-covered

"In typical Nevada fashion, the neighbors placed bets, sitting in their pickups, watching."

Jerrie Tipton spreading hay to get cattle to "impact" a Nevada mine site. After the cows were taken off, the grass grew thicker on this once barren slope than on some of their neighbors' cultivated hayfields.

"…one trip up that slope only gave me the barest hint of what it must have been like to struggle to the edge of endurance to prove a point."

basin-and-range country of which it is a part, Tony and Jerrie Tiptons' ten-acre mine reclamation project seemed infinitesimal. Even the success they have achieved across their entire ranch shrinks into insignificance in comparison to the job that faces anyone with sufficient hubris to think they can improve the West. If hard work ruined all this, can hard work bring it back? Or is this more properly a job for higher powers: Nature? Fate? The federal government?

Tony and Jerrie Tipton consider themselves agents of at least two of those higher powers: nature and fate. When they look at that expanse of decreasing diversity and deepening arroyos, they don't see work in great crushing gobs—they see opportunity, lots of it. You have to be a zealot to do that, and there is more than a little of the zealot in both Tony and Jerrie Tipton. Their house stands as a monument to it and to the hard work it drives them to. Reams of monitoring data and piles of photographs occupy all horizontal surfaces that aren't used for eating or sleeping. Their photographs, especially, offer testimony to an eye for detail and minute evidence of change that borders on the visionary. One photograph that Tony pulls out of a teetering stack looks like an attempt at modern art. It shows a mosaic of golf tees spread more or less evenly across a piece of apparently bare dirt. Tony makes sense out of the photograph for everyone puzzling over it by saying, "See, they're all pointing to seedlings." Looking closer, we all notice the tiny wisps of green off the point of each tee.

Still another photograph shows a scene from what could be a preservationist's worst nightmare.

Sagebrush plants are trampled, tortured, and broken. The bare expanses that separate them are churned to dust. Cow dung is splattered everywhere and scrambled into the devastation. "How's that for animal impact?" Tony asks.

Then he produces another photograph, green and vibrant, with sagebrush dispersed among other plants, grasses, and forbs. It's apparent that both photographs were taken in the same place at different times. Pointing to the tall-grass photograph, Tipton proclaims, "And that's the same place one year later!"

In the winter of 1993, the Tiptons took their animal impact show on the road, busting out of the limits of their Toiyabe Mountains ranch and trailing their cattle up the Reese River into the high deserts north of Death Valley. Along the way, they crossed over the allotments of dozens of other ranchers who would rather stay at home, holding on, waiting for fate, while it may have been walking across their ranch.

For 150 miles and fourteen days, the Tiptons drove their cattle the old way, into an area they call Candelaria, for the ghost-town mining camp located there. Less than five inches of rain a year is the average here, where the grass plants are so far apart someone once joked that you have to cross your cattle with mustangs so they're fast enough to get from one plant to another before they starve to death. Through one of the coldest winters in recent memory, the Tiptons day-herded their animals and hauled water to them, fighting to keep both from freezing.

This is where we ended up on the last day of the team meeting, monitoring the results of the Tiptons' range-improvement road show. On the two days before, back near the mine site, we had visited a sagebrush flat where sedges were beginning to grow in an area that only Tony realized was once a riparian area. We had viewed ephemeral streams that showed signs of running year-round again, as they had when they drove waterwheels in the nineteenth century. We had set up stream-profile monitoring transects to gauge the rate at which deeply eroded creekbeds were filling up with sediment filtered out of floodwater by new-grown riparian vegetation.

All these beneficial changes, Tony and Jerrie said, were the results of their management. Since rainfall was the same on areas that hadn't improved, and natural conditions, therefore, couldn't be credited, no one argued with them—although there were still plenty of skeptics.

At Candelaria, the Tiptons' presence seemed to inspire bad weather, or good weather if you're interested in growing things. Shortly after we arrived, it began to rain and the temperature dropped. "I've never been here when I ain't been cold," Jerrie said. In spite of the weather, she brought out the monitoring tools and we started throwing darts to select random data points in the desert immensity.

The Tiptons use a system of monitoring that many mainstream range biologists discredit as inexact and overly subjective. To them, it is more valuable than traditional monitoring methods because you don't have to be a scientist to use it and because it collects information they find useful

"To them, plants and animals don't compete, they respond to opportunities."

Sagebrush that had become a stagnant monoculture (top photo) has been trampled by the Tiptons' cattle. The soil has thus been opened to the sun (bottom photo) and enriched with sagebrush litter, hay, dung, and urine.

in making management decisions. But even more than that, it reflects their philosophy about nature.

Conventional range science describes rangeland in terms that sound more like a *Star Trek* characters list than a basis for unbiased observation. Plants are pigeonholed as increasers and decreasers, invaders and exotics. The predominant interaction in this scenario is a dog-eat-dog, survival-of-the-fittest competition for scarce resources, a fight to the finish. It sounds like it's more about politics than nature, because it is. The theory on which this scenario is based, of ecological succession and climax communities, was developed in the nineteenth century. It was based on Social Darwinism, a Victorian political theory that held that all cultures were tending toward, were eventually destined to become more like, Western European industrial society. Ironically, while Social Darwinism is officially scorned as a socio-political theory—considered racist among other things—its derivative in the world of ecology still enjoys wide acceptance and respectability.

The Tiptons see range ecosystems in an entirely different way. They see them as a web of cycles instead of struggles. To them, plants and animals don't compete, they respond to opportunities. When conditions are right for a certain plant or a community of plants and animals to grow, it will. When those conditions change or are changed, not only will the living community change but so will the soil and the water and even the landscape.

By changing the way their land is managed, the Tiptons have been able to decrease the expanses of bare dirt that separate living plants there by as much as 80 percent. Streams flow for more days out of the year now, and vegetation is beginning to once again crowd their riparian areas. This has happened during a time when there has been a decrease in precipitation along with increase in grazing pressure. If competition dictated the condition and density of plants, that density should have gone down, but it has gone up. There is, of course, a limit to this, but Tony Tipton says we have no way of knowing what that limit is. "We don't have any idea of what this land's potential is," he says.

At Candelaria, the data we gather will end up being little more than a curiosity. The Tiptons' efforts to revive the land so stimulated the native Indian rice-grass plants that grow there that the rancher who held the allotment permit decided to graze it with his own cattle the next year.

In the meantime, the Tiptons are reacting in typical fashion, moving on to other challenges, reclaiming an even bigger mine site (130 acres), and restoring an entire watershed on another ranch, which will replace the winter range that they lost at Candelaria. In their dedication and their doggedness, Tony and Jerrie Tipton are reminiscent of another rebel who once changed the face of the West. To draw a comparison at this point might be the best way to put the work and dreams of the ranchers and nonranchers who have been described in this book into perspective.

More than any other individual, Major John Wesley Powell, a one-armed Civil War veteran, ex-teacher, explorer, and first head of the U.S. Geological Survey, is credited with making the West what it is today: a mosaic of public and private ownership where much of the land and resources remain in public hands. Though that is the same West that Sagebrush Rebel Tony Tipton tried to tear down, there is another part of Powell's vision that Tipton and others like him (many of whom have never even heard of Powell, let alone his vision for the West) are working to bring to reality.

In "The Legacy of John Wesley Powell," a paper presented at "The Second Opening of the West," a symposium held at the Sharlot Hall Museum in Prescott, Arizona, in November 1994, author and historian Donald Worster told us that, although Powell firmly believed that most of the West should remain in federal hands, he was concerned that any centralized bureaucracy set up to administer the public domain would turn into "a hotbed of corruption." The most effective way to insure that those lands would be well managed, Powell believed, was to empower the people who lived among them to plan the use and conservation of their resources. By reaping the benefits of their successes and suffering the consequences of their failures, Powell believed these empowered citizens would bring to that land the most reliable incentive to manage it well. But this old explorer, who had seen a lot of desert, also knew that the arid West was not the proper environment for the Jeffersonian ideal of a landscape of small farms. Here, he knew it could take thousands of acres instead of the allotted homestead of 160 acres to support a family. Powell's idea, then, was to have the government set up a system of

"commonwealths" situated within major watersheds and inclusive of what we today would call whole ecosystems. Within those commonwealths, residents would own the lands with enough water for irrigation, and the rest would be held in common.

"I say to the government: Hands off!" Powell told a seminar of politicians and private citizens as he described his vision to them in 1890. "Furnish the people with institutions of justice and let them do the work for themselves."

Powell's vision of a self-reliant West managed by its residents was rejected in favor of an idea advanced by the first head of the newly formed forest service, Gifford Pinchot. President Theodore Roosevelt chose Pinchot's idea of a land-management bureaucracy run by idealistic young technocrats over Powell's commonwealths, and that model still drives the management of western public lands.

But Powell's dream might not be dead after all. If you look closely at the TWWMT, there is a lot of the old explorer's vision in it. Although most of the team members may not own the land they're helping to manage, they have taken ownership in the goals they have helped set for that land. The fact that many of them come from well outside the boundaries of the ranch is merely a reflection of the broader sense of ownership of public lands that has become a political reality in our contemporary society.

Powell retired a few years after his proposal for a West of sustainable rural communities was rejected, and he died not too long after that. I don't know what the former Union major would think of a states' rights radical like Tony Tipton taking up his banner and running with it, but I'll bet somewhere in his soul the self-reliant old adventurer would be rooting for him.

"By reaping the benefits of their successes and suffering the consequences of their failures…these empowered citizens would bring to that land the most reliable incentive to manage it well."

EPILOGUE

THE WORK of the Grand Canyon Trust, the organization that sponsored this book, centers on the Colorado Plateau. That land of high deserts, rugged mountains, and some of the most spectacular cliffs and canyons in the world covers most of southern Utah and northern Arizona and extends into southwestern Colorado and northwestern New Mexico. If you know your geography, you've already realized that no ranches on the plateau are included in this book. Why this is so is a matter that deserves some discussion.

For one thing, the omission is not intentional. We didn't set out to produce a book about ranches that are doing a good job, and then choose not to include any from the plateau; nor does it mean that there are no well-managed ranches there. Several were nominated, and for good reason. The fact that we ended up not selecting any came about more as a matter of the way this book was put together than anything else. For a ranch to be suitable to be included here, it had to satisfy three criteria. First, a significant portion of it had to be in a condition that both ranchers and environmentalists could love: diverse, healthy, beautiful, productive, and getting more so. Second, this condition had to be the result of good management. Third and last, but certainly not least, the success of that management had to be the result of collaboration and teamwork between the rancher and other interested parties, including environmentalists and government land-management agency experts.

When we contacted the Colorado Plateau ranches that had been called to our attention, the managers of each outfit said he or she didn't feel their operation lived up to all of the selection criteria. The major reasons given were that they hadn't put forth enough of a collaborative effort, they didn't have enough documentation to prove they had improved the land, or they didn't think their achievement was worth being singled out as one of only ten ranches in the entire West. A few of them just didn't want to be put in that much of a spotlight. Nevertheless, all of the ranchers we contacted were proud of what they were doing, and the people who nominated them certainly thought they were worthy of being profiled.

In the year and a half that has passed since the Trust and I started this book, this situation has changed. As I write these last few pages, there are some new and encouraging developments underway in a variety of areas on the plateau.

A group of ranchers and environmentalists working together in and around the Colorado Plateau communities of Crested Butte and Gunnison, Colorado, so caught the attention of Secretary of the Interior Bruce Babbitt that he used them as the model for the local control and community involvement provisions of his Rangeland Reform '94 proposal. What caught Babbitt's eye was the way the members of the High Country Citizen's Alliance (HCCA), a 350-member environmental group based in Crested Butte, and the Gunnison County Cattlegrowers had joined forces after discovering that they shared a common desire to protect their community from the runaway suburbanization that was gobbling up the open space they all loved so much. As part of that discovery, the members of the HCCA realized that working to promote a sustainable rural economy based on open space—dependent businesses such as ranching was in their best interest because, in this case at least, it was in the environment's best interest.

Along the southern reaches of the Colorado Plateau, near Flagstaff, Arizona, two ranches have formed a collaborative management team dubbed the Diablo Trust. The trust is similar to the management teams formed by several of the ranches profiled in this book (the Orme Ranch, the Tipton Ranch, the Milton Ranch), but in this case, there are a couple of important differences. First, by combining the management planning processes of two ranches, the Diablo Trust gives its members the opportunity to have an impact on a larger landscape than if they were participating on just one ranch. The ranchers involved also end up having to attract fewer people to form a critical mass of team members. By combining their lands, the Diablo Trust ranchers are hoping eventually to be able to combine their herds and grazing rotations in such a way that they can graze areas as infrequently as every second year. The payoff for environmentalists is the promise of more land being covered for longer periods of time with the tall waving grass we usually associate with preserves and park lands rather than ranches. The payoff for ranchers comes in the more concentrated impact, along with its associated benefits, that results from grazing land with a larger herd, and the land gets a longer rest in the process. Norm Wallen, a member of the local Sierra Club's executive committee who has attended several of the Diablo Trust's team meetings, is intrigued by the process. "Though I'm still not sure that what the ranchers say they can do is possible, my impression is that this is a group of people trying to get past all this confrontation crap and do what's best for the land. I'm very impressed."

On the Babbitt family ranch, the CO Bar, which is also located near Flagstaff, ranch managers have joined with environmentalists associated with the 6-6 group (mentioned in Chapter One) to create a study plot on which the results of some of the paradigm-shifting management techniques described in this book can be tried and monitored. That five-acre plot is bordered on one side by Wupatki National Monument, which has been rested from livestock grazing for six years, and on the other three sides by ranchlands grazed as they have been for more than a century. On the study area, extremely short-term grazing episodes are being tried at animal densities up to twenty-six times as high as is recommended by conventional management for rangeland in this area. In spite of that intense impact (400 animals on five acres for just short of one day), plant spacings on the study plot have remained the same as they are on the two control plots, and the vitality of native grasses has improved considerably on the study area while remaining the same on the other two areas. Though the study process is only in its second year and results are still too preliminary to support any hard conclusions, they have remained consistent through two growing seasons: one very wet, the other very dry.

Heidi Redd owns the historic Dugout Ranch, which borders the Needles District of Canyonlands National Park south of Moab in southeastern Utah. Recently, she and the Utah Division of Wildlife agreed to cooperate on a five-year monitoring project in an area named Beef Basin near the park, where the vegetative community has deteriorated enough to cause concern. "To a casual observer, it's the deer that are causing the damage," explains Heidi. "And, some say it's been the cows. I'm not in it to fix blame, but to find out how to make the range better."

Since people have to drive across her ranch to get to several of the more popular areas of the park, Redd says she has set a goal to improve her land to the point that people enroute can't tell whether they are on park land or ranch land. "Sometimes, especially when it's dry, that's not all that easy to do," she admits. "The cows have got to eat, so I've got to move them around. To see how the land looks then really tugs at my heart." In some areas of the ranch, including the climbing area, Redd has stopped grazing cattle altogether to keep her land visually compatible with park land.

Heidi has also struck up a working relationship with rock climbers, with whom she had been at odds, to repair damage resulting from overuse of popular climbing areas on private lands near the park. Because of trampling, off-road driving, and littering, Redd was ready to close down one area that includes a world-famous climb named Super Crack and to discontinue access across the ranch to other climbing areas on public land. All this was averted, however, when she and the climbers came to a mutually beneficial agreement. Redd agreed to keep climbing areas and access routes open if the climbers would repair some of the environmental damage that had been done and if they would do something to stop the proliferation of roads and trails into the area that had denuded large areas of land. "They blocked off some of the roads and gathered native grass seeds off the ranch and planted them in places that were bare," Redd says. "When we work together, it's amazing what we can accomplish. When we fight, nothing gets done."

One of the most unusual grazing-related restoration projects on the Colorado Plateau involves rancher and writer Steve Rich. Rich's family owns a ranch that includes three public lands grazing allotments on the Arizona Strip. This wild and remote stretch of the plateau is bordered on the south by the Grand Canyon and on the north by the Utah state line. The Strip has been the site of some of the most celebrated environmental abuses associated with livestock grazing in the Southwest. Overgrazing has been blamed for creating such devastated conditions on Rich's family's 50,000-acre ranch that the ranch at present supports only ninety-six cattle. "In some places there's just nothing for anything to eat. The land is sterile," says Rich ruefully, realizing that it's almost impossible for someone who has never seen such desolation to understand. As were Tony and Jerrie Tipton, however, Rich is challenged rather than paralyzed by such apparently hopeless conditions. In order to determine if this land is capable of sustaining grazing, he says he is asking the sandy soil and the plants that grow on it if they remember the interrelationships that existed between grasses and grazers in the days when the Pleistocene megafauna lived here. Can plants and dirt remember something that happened more than 11,000 years ago? Biologists tell us they can. In terms of evolution, 11,000 years is barely the blink of an eye.

In a collections room in the Colorado Plateau Research Station on the Northern Arizona University campus in Flagstaff, paleo-ecologists Larry Agenbroad and Jim Mead of the Quaternary Studies Department tell me that, a blink of an evolutionary eye ago, what is now sand and slickrock on the Colorado Plateau "must have been incredibly rich to support mammoth,

camels, horses, bison, shrub oxen, and extinct mountain goats." Mead and Agenbroad have located dung deposits on the plateau that still have traces of these animals' digestive bile in them. The bile can be analyzed to tell the species of the animal from which it came. Using information gathered in this and other ways, Agenbroad is in the process of redrawing the map of bison distribution across the West. Before Mead and Agenbroad made their discoveries, bison were considered to have been absent from most of the Southwest, including parts of southern Utah and almost all of Arizona, in any time frame that could be considered to be relevant to current ecosystems. Agenbroad has found bison bones and ancient Indian etched rock art (petroglyphs) of bison-hunting scenes in areas where the grasses supposedly evolved without the presence of large grazers. Some of those finds spill over into eastern California, eastern Nevada, and western New Mexico. Some are relatively recent, dating back to the period just prior to the arrival of Europeans.

From mummified dung specimens, Mead and Agenbroad can tell what species of plants these ancient animals ate when they lived on the plateau. Poking around, armed with that information, the two scientists have concluded, in Agenbroad's words, that "some of that habitat is still with us." They have found it, he says, in relict grasslands hiding out on shaded slopes and on isolated meadows 4,000 feet higher than where the dung heaps were located. These are the habitats Steve Rich is attempting to revive on the Arizona Strip and to spread into areas that now raise little more than sagebrush and an occasional cholla cactus.

To ask the land if it can remember its once-prolific ways, Rich concentrates his cattle so they graze as a herd of ancient bison might have, and then he watches to see how the ecosystem responds. The technique sounds odd, but it has its scientific precedents. In his essay, "Nature's Odd Couples," in the collection *The Panda's Thumb*, Stephen Jay Gould writes of biologists using a similar technique—feeding the seeds of a large tree, Calveria major, to domestic turkeys—to see if the tree's population decline on its home island of Mauritius could be due to a now-unrequited dependence on an extinct cohabitant, the dodo bird. The seeds were fed to turkeys, which are about the same size as dodos, to see if the birds' gizzards would abrade the nearly impenetrable seed shells so they would germinate. If they did, that would provide evidence that dodo gizzards may have performed the same favor for the trees. The turkeys were able to turn out the first Calveria seeds known to have germinated in 300 years, but the verdict was still considered inconclusive because, well, only a dodo could answer that.

In a similar instance, scientists considered that Central American trees with seeds that seem to be ineffective dispersers might be missing the mastodons that once most likely consumed their seeds and dispersed them. When it was found that other large browsers (cattle and horses) could enhance dispersal of those seeds, the answer arrived at was, most likely they do.

Has Steve Rich's experiment told him anything? He says it has. In some areas, the land has responded by growing as many as 100 perennial grassplant seedlings per square foot. Although other areas have not responded as well—"There's no organic material left in the soil there," says Rich—he is still encouraged enough to plan to turn grazing into an art medium that involves using his herd as a huge brush, eating and trampling its way across the landscape. He hopes the result will be a green, growing swath captured in time-lapse photography.

Rich, the BLM, and the forest service are working on a management plan for his ranch's three allotments that will take into consideration the style of management that created those plots brimming with seedlings. They are also working to pull together collections of ranches to create planning areas that more nearly approximate the boundaries of ecosystems, which federal agencies are now required to manage.

"We're still trying to define an ecosystem," says Bob Sandberg, branch chief for renewable resources in the BLM's Vermilion Resource Area on the Arizona Strip. "What we've decided is that we have to look at the land holistically, even if the whole is as small as a single allotment. Then we're working to bring all the interested parties together to come up with common goals and work toward them." Inspired by a talk by conflict resolution facilitator Tommie Martin, the BLM is putting together teams for a number of areas on the Strip. Their hopes are high. "We've become so tied up in paperwork and controversy, I can see us getting almost immobilized. We're hoping this type of concept will break us through," says Sandberg, with a touch of optimism.

To Steve Rich, all this adds up to a fascinating detective story and a way to challenge his creativity. To Bob Sandberg, it promises a way to stop dealing almost entirely with controversy and to start dealing once again with the land. For others of us, all it does is complicate matters. "When I think of the Colorado Plateau, I think of it as sandy soil with cryptobiotic crusts on it," said Rick Moore of the Grand Canyon Trust, with whom I've been working on this project. "The plateau is as much that to me as it is red rock cliffs and Anasazi ruins. I don't know if I want cows and cowshit and grass all over the place, even if they can grow it there. And I'm pretty sure I know a lot of other people who don't want it that way either."

Poets describe the plateau's beauty with words such as "spare" and "stark." Can spare beauty survive animal impact? What if cows can get grass to grow on the plateau? In southern Arizona they're growing it on bare mine tailings. What do we do? How do we decide?

"All decisions must be made on the basis of the best science," read an advertisement published by a consortium of environmental groups in dozens of newspapers around the country. The only problem is, both sides have their own "best science." Parks and preserves typically determine the health of grasslands by measuring the amount of vegetative canopy (the amount of standing plant material). Range biologists associated with ranches, on the other hand, record live basal area (the total area occupied by living plants).

While monitoring an area of high desert on the Colorado Plateau near Flagstaff, I found that more than half of the plants within the monitoring transect were seedlings of native bunchgrasses. Not only was the amount of ground covered by live plants relatively high on this piece of land but the vitality of the plant community in terms of how many "babies" it was having was enormous. Since this area had been heavily grazed for three weeks prior to being monitored, the amount of standing vegetative material on it was insignificant compared with a nearby ungrazed site, where, after six years of rest, two-thirds of the plants were stagnant or dying and no seedlings were found at 100 monitoring points. If you measured live basal area, the grazed area was the healthiest. If you measured vegetative canopy, the ungrazed area scored far and away the highest. Each of those two pieces of land was in excellent or horrible shape depending on which study you used to evaluate it. (And you wondered how something that sounds so simple could be so complex.)

Both of the studies, of course, are valid. And both give us important information about the land. The problem arises when competing factions cite one and ignore the other as proof that their prescriptions for the land are based on "best science." Neither science nor the land tells us whether it is better to have a high percentage of seedlings and short green grass, or to have tall, dry grass waving beautifully in the breeze, or communities of cryptobiotics on otherwise bare dirt. Most animals—wild and domestic, big and small—prefer to eat green plants. I guess that tells us their vote. As for which of these conditions we should choose to manage toward, that's for us to decide.

At this point, rancher Ivan Aguirre would ask, "What is your goal?" And we must realize that even not to choose is to make a choice: a choice to abandon the land, and therefore ourselves, to the still-unfolding results of the mistakes of the past.

Perhaps the most important thing the people profiled in this book can teach us is that we can choose to have rural communities in the West with sustainable economies based, at least in part, on a vested interest in the ecosystems that surround them, in having those ecosystems healthy and diverse, in restoring what elements of them have been lost. To the citizens of these communities, that would mean regaining a sense of pride in their past and in a lifestyle they have already shown they will go to war to protect. To all of us, it means having a West populated by communities with an economic base connected to, and mindful of, the carrying capacity of the land. In such a West, there would be less reason for communities to turn themselves into theme-park caricatures to attract tourists, or to tweak the tiger's tail of urbanization and open-ended growth by courting "low-impact" industry. In such a West, we could live more as a part of the ecosystem and less as a liability upon it.

Environmentalist, forest ranger, and author Aldo Leopold wrote in his essay "Wilderness," (in *The River of the Mother of God and Other Essays*, University of Wisconsin Press, 1991): "The impulse to save wild remnants is always, I think, the forerunner of the more important task of mixing a degree of wildness with utility." To

Leopold, achieving that mix meant using "a positive exercise of skill and insight, not merely a negative exercise of abstinence and caution." The result, he believed, would be "a state of harmony between men [and women] and the land." Perhaps we could add the words, "and between one another," to that last quote.

Eldridge Cleaver helped define the era to which most of us trace our political coming of age by saying there is no such thing as common ground. "You're either part of the problem or part of the solution," he told us. When it comes to the environment, Cleaver was wrong. We're all part of the problem, so we all must be part of the solution.

NOTES ON THE PHOTOGRAPHS

THE PORTRAITS were made with either a 4x5-inch Linhof Technika 3, using Kodak Tri-X Film-Pack, or a 5x7-inch K. B. Canham view camera, using Ilford HP5+ film. Lenses ranging in focal length from 90 to 240 millimeters were used for the portraits. The Gray Ranch landscape on the title page was photographed with my homemade 4x10-inch panoramic camera with a 90mm Super Angulon XL lens, using Ilford FP4+ film and a red-orange (23A) filter. My wooden Ries C-100 tripod, equipped with a Linhof 3663 head, provided solid support on all setups.

For determining exposures I used a Pentax V spotmeter and employed the Ansel Adams Zone System for coordinating exposure and development. The Tri-X film was developed in Kodack HC 110 (1:31), the HP5+ in Zonal Pro Gamma Plus (1:12), and the FP4+ in Kodak D 76 (1:1).

The reproduction prints were made on gloss-surface Forte fiber-base papers: Fortezo (graded) and Warmtone Polygrade. These Hungarian chlorobromide papers were chosen for their magnificent tonal scale and their exceptional responsiveness to development controls, reducing, and toning. They were developed in either Edwal 106 (1:15) or Zonal Pro HQ Warm Tone (1:10). For between-grades split development I used Zonal Pro Vara Soft (1:15 or 1:20) as the first developer and Edwal 106 as the second.

The portrait prints were made with a Durst 5x7 Laborator enlarger, using either the condenser or the cold-light mode. The landscapes were printed with an Elwood 8x10 enlarger equipped with a cold-light head.

Selective reduction (bleaching) with a solution of potassium ferricyanide and sodium thiosulfate was done on most of the prints. An integral step in my printing procedure, the reducing was used to achieve degrees of tonal adjustment and balance too subtle for the conventional exposure controls of dodging, burning, and flashing that I also employed extensively. The prints were toned in Kodak Rapid Selenium (1:31).

J.D.

AFTERWORD

IN THREE DECADES of prowling the American West with camera and saddle, I have come to cherish the ranching way of life and the opportunities I have had to be a part of it. I particularly admire the art of "cowboying," the finest moments of which involve that working partnership forged between *Homo sapiens* and *Equus caballus*—an artistry that harks back to the ancient interdependency of two-leggeds, four-leggeds, grass, sun, and rain that we call pastoralism.

As I visited the ranches that my partner, Dan Dagget, has profiled in this book, I inevitably inquired as to the role of cowboying in their present-day scheme of things. In Wyoming, Jack Turnell informed me, proudly, that they rope their calves for branding at the Pitchfork Ranch, and that, in keeping with the long-standing western custom of "neighboring," help comes in from, and goes back out to, the ranching community at large. At the Sieben, in Montana, Chase Hibbard prefers that ranch work be done on horseback whenever possible. Jerrie and Tony Tipton of Big Creek Ranch, in central Nevada, are on horseback with their crew for weeks at a time, slow-trailing a large herd in order to accomplish planned rotations and provide "herd effect" land treatment along the route. In southwestern New Mexico, Brother Dagget and I spent two early spring days on Gray Ranch horses, helping Drum Hadley and his cowboys gather and brand calves. It was a chance once again to experience the West at its best—the West as I envision it, the West where cowboying remains absolutely in context.

My vision for the West runs like this: Crystalline air—that salubrious medium that so perfectly displays fabulous sculptural landforms bathed in glorious light; monumental clouds that are strictly weather, not blobs and smears that are emissions; pure water— flowing again where it used to, running wild where it should; biodiversity in forest, meadow, desert, and plain; wildlife in abundance; roadbuilding and "leapfrog" develop- ment in check; and habitat preserved. In general, less people. In particular, more people caring, communicating, collaborating, conserving. Rural economies that work, and wilderness that works.

A "working wilderness," if you will. One that, in a healthy and balanced way, sustains wildlife and ecosystems, livestock and other productive uses—and itself. One that works, not in some entrepreneurial sense that everything must conform to man's idea of utilitarianism, but because working is what it's always done, and because working is what keeps it free. One that doesn't proclaim itself WILDERNESS in flashing neon to attract the many, but one that excludes no individual who truly appreciates and nurtures it.

Achievable? Who knows? Worth working together for? You bet!

JAY DUSARD, *Miller Creek, Arizona, October 1994*

DAN DAGGET   JAY DUSARD

DAN DAGGET has been involved in environmental issues since the 1970s, and was honored by the Sierra Club in 1992 as one of the top grassroots activists in America. His articles have appeared in *Outside, Sierra,* the *High Country News,* and other magazines and newspapers. In addition to being a writer, he is a biological consultant, public speaker, and collaborative team-building facilitator.

JAY DUSARD was raised on an Illinois farm, studied architecture in Florida, cowboyed on the Arizona–Sonora border, and taught photography at Prescott College in Arizona. A 1981 Guggenheim Fellowship took him, with camera and saddle, from British Columbia to Chihuahua and resulted in his acclaimed first book, *The North American Cowboy: A Portrait.* While working on *Beyond the Rangeland Conflict,* Jay also completed *Open Country,* a 28-year collection of panoramic landscapes celebrating the West as he is dedicated to keeping it—open, wild, and free.